Conceptual Meteorology

NIPA® GENX ELECTRONIC RESOURCES & SOLUTIONS P. LTD.
New Delhi-110 034

About the Authors

BV Ramana Rao (b 1940) taught Agricultural Meteorology in the University of Agricultural Sciences, Bangalore during the years 1966 to 1979 and later joined the Indian Council of Agricultural Research and superannuated from service during April, 2000. He served as Head, Division of Climatology, Wind Power and Solar Energy Utilization at ICAR-Central Arid Zone Research Institute, Jodhpur (Rajasthan); founder Project Coordinator, All India Coordinated Research Project on Agrometeorology at ICAR-Central Research Institute for Dryland Agriculture, Hyderabad (Telangana) during December 1984 to March 1994. His major contributions in the field of Agrometeorology include Agroclimatic Characterization, Delineation of Efficient Cropping Zones, Crop Weather Modeling, Classification and Management of Agricultural Droughts, Operational Agrometeorology, etc. He published about 150 research papers in peer reviewed journals with 50 of his papers appearing in high impact International Journals. He authored 5 books and monographs and a number of scientific and technical bulletins. He served as Editor in Chief, Journal of Agrometeorology during the years 2009 to 2023.

He is the recipient of British Council Fellowship during the year 1986-87 for advanced training at University of Reading, United Kingdom. He was conferred as Honorary Fellow, Association of Agrometeorologists during the year 2011 for his valuable contribution to the growth and development of Agricultural Meteorology in India.

Surender Singh (b 1965) has about thirty five years of experience with PhD in the field of Agrometeorology. Worked in Agrometeorology encompassing teaching, research, extension and administration at various capacities viz., Principal Scientist, Professor & Head, Senior Scientist, Assistant Scientist, Advisor (Recruitments) and Associate Director (Students' Counseling and Placement) at CCS Haryana Agricultural University, Hisar, India.

He is life member of Association of Agrometeorologists, Indian Meteorological Society, Indian Society of Remote Sensing, International Society for Agrometeorology, Society for Sustainable Agric & Resource Management, Environment Science Published for Everybody Round the Earth (ESPERE), European Drought Centre, South Asian Forum on Agriculture Meteorology (SAFOAM) and The Geographical Society of India. Organized quite a few trainings/symposia in Agrometeorology and associated domains at National level. Published more than 250 research papers, popular articles, manuals/teaching aids and research bulletins/book chapters etc. Performed responsibilities as Managing Editor-Journal of Agrometeorology, Research Collaborator, Monsoon Asia Agro-Environmental Research Consortium (MARCO), Japan, Member, Expert Team-CAgM-WMO, Geneva, Switzerland (2018-20). Core Committee Member on Agriculture-Drought at WMO, Geneva, Switzerland (2021-24 & 2024-27). External Expert for evaluation of research projects submitted at COST-Association, European Union, Brussels, Belgium. Many students (UG, PG and PhD) were trained under him spread across the country and settled in various positions in the field of Agrometeorology. Widely travelled across 15 countries on academic pursuits.

V Uma Maheswara Rao (b 1954) has forty years of experience with PhD in the broader field of Agrometeorology. Worked in Agrometeorology encompassing teaching, research, extension and administration at various capacities *viz.*, Project Coordinator, Principal Scientist, Professor & Head, Agrometeorologist, and Assistant Scientist at CCS Haryana Agricultural University and Indian Council of Agriculture Research in India.

He is life member of Indian Meteorological Society, The Arid Zone Research Association of India, Indian Society of Remote Sensing, Association of Agrometeorologists, International Society for Agrometeorology, and The Indian Society of Dryland Agriculture. Organized several trainings/workshops in Agrometeorology at the International and National level. Published 344 Research papers, popular articles, manuals/teaching aids and research bulletins/book chapters. Member of the CCI-ET-IIC, WMO, Geneva (2014-2017). Member of XI plan High Power Review Committee of Ministry of Earth Sciences, Government of India (2012). The prestigious Choudary Devi Lal Award, ICAR best AICRPAM Centre was obtained under his dynamic leadership as project coordinator in 2014. Many students (UG, PG and Ph.D) were trained under him spread across the country and settled in various positions in the field of Agrometeorology. He visited 10 countries in his official capacity.

Conceptual Meteorology
For Master's Students in Agricultural Meteorology

BV Ramana Rao
MSc Tech; FAAM
Project Coordinator-Rtd (Agrometeorology)
ICAR CRIDA, Hyderabad, Telangana, India

Surender Singh
MSc; PhD
Professor / Principal Scientist (Agrometeorology)
CCS HAU, Hisar, Haryana, India

V Uma Maheswara Rao
MSc; PGDIAS; PhD; FAAM
Project Coordinator-Rtd (Agrometeorology)
ICAR CRIDA, Hyderabad, Telangana, India

NIPA® GENX ELECTRONIC RESOURCES & SOLUTIONS P. LTD.
New Delhi-110 034

NIPA® GENX ELECTRONIC RESOURCES & SOLUTIONS P. LTD.

101,103, Vikas Surya Plaza, CU Block
L.S.C. Market, Pitam Pura, New Delhi-110 034
Ph : +91-11-43860225, Mob.: +91 9717133558, 9540816132
E-mail: newindiapublishingagency@gmail.com
Website: www.nipaersources.com

Print ISBN: 978-93-58874-50-1

ebook ISBN: 978-93-58871-89-0

Composed and Designed by NIPA®.

Rupa Kumar Kolli, Ph.D.
Honorary Scientist & Former Executive Director
International Monsoons Project Office (IMPO), IITM, Pune
Former Chief
World Climate Applications & Services Division, WMO, Geneva
Former President
Indian Meteorological Society, New Delhi

Foreword

Since time immemorial, human civilizations have evolved by observing their environment, understanding the patterns and phenomena that govern the different environmental components and their inter-relationships. The sky has always been a source of wonder, mystery and inspiration, and this curiosity has fuelled the development of meteorology, the science that not only seeks to predict the weather but also to unravel the intricate dance of forces shaping our climate. While this book starts with the ancient reference to Aristotle, the role of meteorology in human evolution goes back to the earliest times, and indeed there is anthropological evidence that climate played a crucial role in shaping human brain evolution, favoring adaptability and cognitive development.

Meteorology, the science of the atmosphere, holds a unique position in our lives. It affects our daily activities, shapes our environments, and influences long-term climate patterns that impact the future of our planet. Despite its profound importance, the intricate workings of meteorological phenomena often remain shrouded in complexity, accessible only to those with specialized training.

"Conceptual Meteorology" is a journey into this dynamic field, presenting the foundational concepts and principles that form the bedrock of our understanding of atmospheric processes. This book aims to bridge the gap between complex scientific theories and practical understanding, making the subject accessible to students, enthusiasts, and professionals alike, particularly in relation to agriculture, providing a foundational textbook in the field of agricultural meteorology.

As we delve into the chapters that follow, we will encounter a comprehensive yet easily understandable exploration of key topics. 'Conceptual Meteorology' is structured to provide a clear and logical progression through the fundamental concepts of atmospheric science. We begin with an exploration of the Earth's atmosphere, examining its physical processes and the parameters and laws that dictate its behaviour. From there, we delve into the dynamics of weather systems, the formation of clouds and precipitation, and the complex interactions that lead to severe weather/climate events. The final chapters of the book address the characteristic features of Indian weather and climate, including aspects of global teleconnections and climate change. Each concept is presented with clarity and simplicity, supported by illustrations and real-world examples that highlight its relevance and application. The book's approach emphasizes the interconnectedness of meteorological phenomena, encouraging readers to appreciate the holistic nature of the Earth's atmosphere, on aspects that are directly relevant to agricultural applications in the Indian context.

In writing this foreword, I am reminded of the profound impact that a deep understanding of meteorology can have, particularly in the field of agriculture and allied disciplines. Whether one is a budding meteorologist, a seasoned professional, or simply someone with a keen interest in the weather, this book will equip him/her with the knowledge and insights needed to navigate the ever-changing skies.

"Conceptual Meteorology" is more than just a textbook; it is an invitation to see the world of agriculture through the lens of a meteorologist. It challenges one to think critically, to question assumptions, and to embrace the complexity and beauty of our planet's atmosphere. As the reader embarks on this intellectual adventure, I hope he/she finds not only answers but also new questions that inspire further exploration and discovery.

I commend the authors for their dedication to making meteorology accessible and engaging to the postgraduate students of agricultural meteorology in India. I have known all the three authors over the past more than four decades and can vouch for their passionate pursuit of knowledge in support of agricultural meteorology, and their lifetime achievements in building bridges between meteorology and agricultural sciences. Their work is a testament to the importance of science communication and education in fostering a deeper understanding of the weather and climate sciences, and their potential applications in the oldest socio-economic sector of agriculture.

Rupa Kumar Kolli

Rupa Kumar Kolli

Preface

The National Commission on Agriculture, duly constituted by Government of India, in their report Volume IV on Weather and Agriculture (1976) recommended that all the State Agricultural Universities (SAUs) should establish a separate Department of Agricultural Meteorology and start teaching program in Agricultural Meteorology at Post Graduate level to generate well trained manpower to undertake research in Agricultural Meteorology for better understanding the effects of weather and climate in planning and management of agricultural production which is very much necessary for guiding farmers in carrying out field operations based on the information continuously generated by the National Meteorological and Hydrological Services. As a consequence few of the very progressive State Agricultural Universities started separate Department of Agricultural Meteorology and introduced MSc and PhD Programs in Agricultural Meteorology from early 1980 onwards . As on now, there are more than 60 Agricultural Universities in the Country and it is unfortunate that there are only 19 SAUs have a separate Division/ Department of Agricultural Meteorology and rest of the Universities have either ignored or unaware of the recommendations of National Commission on Agriculture. The Association of Agrometeorologists in collaboration with the ICAR-Central Research Institute of Dryland Agriculture, Hyderabad organized a Brain Storming Workshop on Reorientation of Education and Research in Agricultural Meteorology in the Context of Climate Change and Agrometeorological Advisory Services during September 29-30, 2023. The workshop has recognized the importance of revising the syllabus followed for MSc and PhD programs taking into consideration the recent technological advances made in weather forecasting, creation of data base, modern methods available for analysis and interpretation of multiple data sets available with wide range spatial and temporal variability. The Workshop also identified lack of uniformity in the Syllabus prescribed by the Indian Council of Agricultural Research and the syllabus followed by the Agricultural Scientists Recruitment Board, as a result of which there was some deficiency in training of students. As a follow up action the Association of Agrometeorologists has constituted a Working Group with one of the present authors, Dr Surender Singh, Professor of

Agricultural Meteorology at CCS Agricultural University, Hisar as Convener to develop Model Syllabus. The Syllabus was released in the Inaugural Session of the International Symposium (INAGMET-2024) organized by the Association of Agrometeorologists (AAM) at Banaras Hindu University, Varanasi on 8th February, 2024. As per the model syllabus, a course AGMET 501 was designed integrating the course contents from AGMET 501 of ICAR and Units I and II prescribed by the Agricultural Scientists Recruitment Board (ASRB), New Delhi. The syllabus covers fundamentals of Physical Meteorology and basics of Climatology which are very much necessary for professional development of the students in the discipline of Agricultural Meteorology.

Usually the students opting for career in Meteorology and Atmospheric Sciences require good background of Mathematics and Physics. In the State Agricultural Universities of India, the students opt for career in Agricultural Meteorology at the end of studying Physics and Mathematics prior to University Education. Therefore there is need to make available books to cater to the needs of aspiring students for gaining insight and understanding of fundamentals and basic concepts easily so that they can acquire further knowledge by going through Advanced Text Books and Reference Material. With this objective, we attempted to write the Present Book on 'Conceptual Meteorology' in its simplest form intended to cater the needs of students doing the Master's in Agricultural Meteorology. We believe that such simple books have to be developed for different courses in Agricultural Meteorology by the Senior Faculties and Learned Scientists to promote aptitude for learning and acquiring better professional approach to serve the needs of Agricultural Sector with better dedication and commitment. The authors are wholly indebted to Dr Rupa Kumar Kolli, Honorary Scientist & Former Executive Director, International Monsoons Project Office (IMPO) at IITM, Pune, former President, Indian Meteorological Society (IMS) and Former Chief, World Climate Applications & Services Division at WMO, Switzerland for his thoughtful message by writing foreword for this book.

There is always scope for improvement and we appreciate suggestions from one and all who have entirely examined the book at hand.

BV Ramana Rao
Surender Singh
V Uma Maheswara Rao

Acknowledgements

The authors are extremely thankful to Dr Sompal Singh, Professor - Agrometeorology Punjab Agricultural University, Ludhiana, India for his pertinent contribution in updating bibliography for the book in hand.

The open source text material/images used in the book are also duefully acknowledged.

Contents

List of Figures

1

Introduction

Aristotle (384 to 322 BC), an ancient Greek philosopher and scientist (Fig. 1.1) developed the concept of meteorology as a study of the lower atmosphere. Over the years meteorology developed as a science of atmosphere which deals with the physics, dynamics and chemistry of the atmosphere and their effect upon the earth's surface, oceans and thereby life in general either directly or indirectly.

Fig. 1.1: Aristotle - Greek Philosopher and Scientist
https://www.britannica.com/biography/Aristotle

As the atmospheric conditions change from time to time in a given area and from place to place in a given time, climatology was developed as a science which studies the average conditions of the weather and its variability over a long period of time.

All most all the human activities are greatly influenced by the weather and climate to a greater extent. Therefore, the scope of meteorology encompasses various activities which are of considerable economic significance as detailed here below;

1.1 Housing

The design and construction of houses, buildings, industrial structures etc. depends upon the weather and climate of the region. Orientation of doors, windows roofs of the buildings are usually designed to provide maximum comfort and convenience to the people.

1.2 Agriculture

Agriculture is the most important activity to meet the requirement of food, fiber, fodder, firewood and timber though crop production, animal husbandry, forestry, horticulture, sericulture and fisheries. The choice of crops, and genetic material and production practices, storage, marketing etc. depends upon the weather and climate of the region.

1.3 Aviation

During recent times, aviation sector has grown enormously for transportation of people and material over longer distances in a short period. The pilots are guided by the atmospheric conditions from starting location to the destination for safer transport. The pilots will be provided with the detailed information on the clouds, temperature, visibility, air movements, thunderstorms and turbulence in the atmosphere throughout its route for safety.

1.4 Navigation

For safe navigation, information on adverse weather conditions like cyclonic storms, tidal waves, strong and gusty winds, and roughness of the sea has to be completely provided to the navigators from time to time.

1.5 Irrigation and Water Resources

Planning, development and management of water resources to meet the water requirements in domestic, agricultural and industrial sectors is carried out taking into consideration all aspects of weather and climate of the region. The size and storage capacity of the water reservoirs is determined based on the water requirements in various sectors of human activity round the year.

1.6 Human and Animal Health

Sudden changes in the atmospheric conditions may at times trigger, spread and intensification of various epidemics and diseases. Some diseases may get aggravated and affect large number of populations simultaneously. Therefore, public health authorities are guided to provide vaccination and other immunological measures to reduce the severity of the diseases and epidemics.

1.7 Industry

The industrial sector depends upon the basic raw material produced by the agriculture sector which is controlled by the weather conditions. Therefore, many of the agro-based industries which depend on raw material through agricultural production depend upon the continuous flow of information on weather and climate of the region.

1.8 Commerce

The needs of the people change from time to time depending upon the prevailing weather conditions during different seasons of the year. Therefore, the material supplies through commercial activity are largely driven by the weather and climate of the region.

Thus, all most all human activities including transport, production, tourism, commerce, industry etc. depend upon the meteorological information and therefore the scope of meteorology is getting more and more enlarged with the progress of time all over the world.

1.9 Importance of Meteorology and Climatology in Agriculture

Weather and climate are natural resources and provide key input in agricultural planning and management as all the biological processes in terms of growth and development of living organisms depend upon weather parameters. In addition, all the management practices including land preparation, choice of genetic material, weed control, fertilizer application, control of pests and diseases, harvest and transport of farm produce have to be carried out depending upon the weather conditions from time to time. Role of climate and its interactive linkages in production system have been depicted in Fig. 1.2.

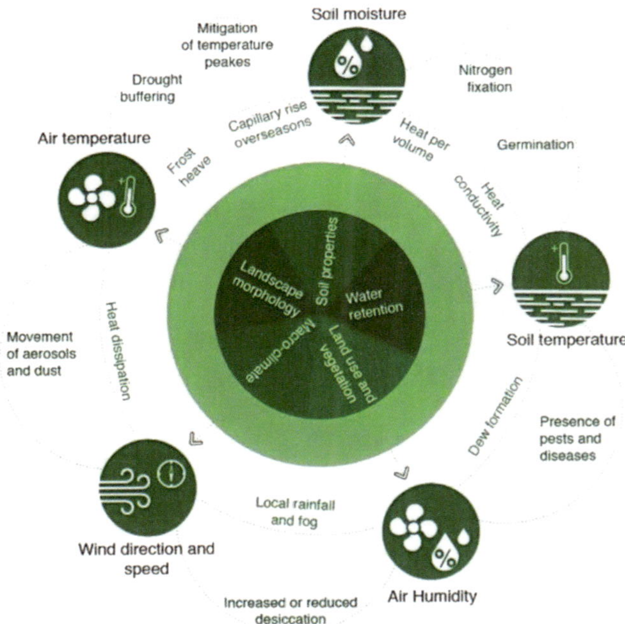

Fig. 1.2: Components of Microclimate and Its Interactive Linkages
https://www.quora.com/What-is-the-relation-between-agriculture-and-weather

Therefore, the nature and type of information required from time to time in agriculture sector is as follows:

1.10 Agroclimatic Characterization

The crop growing seasons have to be delineated based on the information on the nature and type of the soils, depth of the soils, thermal and moisture regimes of a given area or region. In tropical arid and semi-arid regions, water is a severe limiting factor for agricultural production although thermal regime is very favorable round the year. On the contrary, in temperate climates and extra tropical regimes, thermal regime may not be favorable round the year for crop production even when moisture is available in abundance. Therefore, the agroclimatic characterization is prerequisite to determine the extent to which agricultural production is possible in a given region.

1.11 Agricultural Planning for Stable and Sustainable Production

The various crops/ cropping systems that are best suitable for a given region can be decided based on the thermal and moisture regimes depending upon the soil factors. Based on the analysis of long-term weather data, the weather associated risks for growing different crops can be evaluated. The information can be used to identify the choice of crops/varieties which are least vulnerable to changing weather conditions for ensuring stable and sustainable agricultural production in a given region.

1.12 Agricultural Operations Based on Weather Forecasts

Day to day agricultural operations starting from land preparation, sowing of crop, weed management, fertilizer application, and pest disease control measures and harvesting can be scheduled utilizing information provided through daily weather forecasts as well as medium range weather forecasts.

Accurate weather forecasts are very useful for scheduling day to day agricultural operations with greater economy and precision.

1.13 Monitoring Crop Growth and Development

During recent years, considerable efforts were made to develop crop growth models for a wide range of crops. These models can predict the growth and development of a crop on day-to-day basis as driven by weather parameters and management conditions. In the event of adverse weather conditions likely to occur as detailed by the weather forecasts, the impact of such conditions on the performance of the crop can be assessed and the various management options can also be evaluated with the help of decision support systems. Therefore, studies on the effect of weather parameters on growth and development of crops are gaining lot of importance all over the world.

1.14 Biotic Interferences to Crop Growth

Most of the arable crops are very often affected by the occurrence, spread and intensification of pests and diseases (biotic) and abiotic stresses (Fig.1.3). Weather conditions, may sometimes provide very favorable environment for the pests and diseases for multiplication and spread to epidemic proportion. Therefore, studies on the occurrence, spread and intensification of pests and diseases to field crops as affected by the weather parameters is becoming an important area of study.

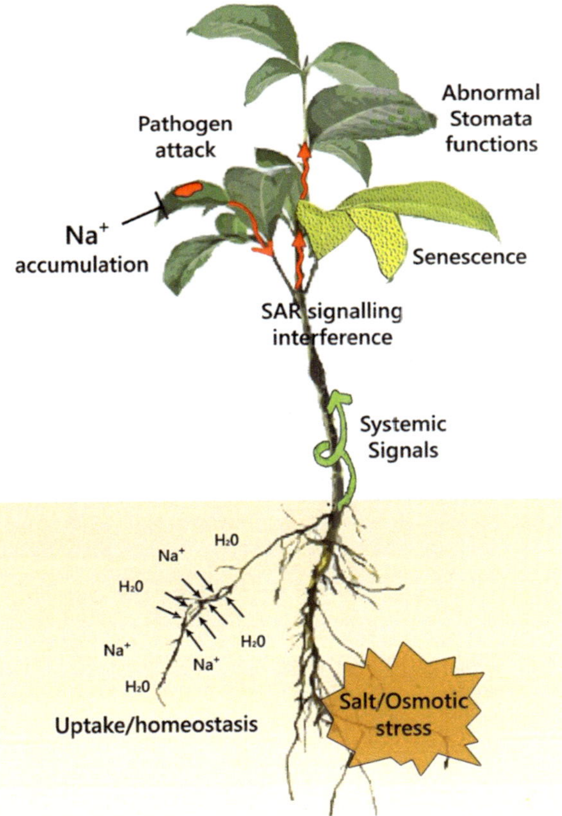

Fig. 1.3: Effects of Abiotic and Biotic Stress on a Plant
https://www.researchgate.net/publication/262932181_Enhancing_crop_resilience_to_combined_abiotic_and_biotic_stress_through the dissection of physiological and molecular crosstalk/figures?lo=1

Even the application of chemicals used for pest and disease control has to be carried out depending upon the suitability of weather conditions at the time of application for their greater affectivity.

1.15 Weather Hazards

Extreme weather conditions like floods, droughts, strong winds, hail storms etc. cause considerable damage to crop production. If the occurrence of such extreme weather events which are detrimental to crop growth and yield can be predicted well in advance, it may be possible to take some protective measures. Information on the occurrence of such events with regards to frequencies and intensities will be of greater use in long term planning and development of management strategies to reduce losses to a certain extent.

1.16 Weather and Fisheries

In coastal regions, several thousands of people depend upon marine fisheries as their main occupation. These fishermen have to be guided daily from time to time about the weather conditions as well as roughness of the sea in coastal regions to ensure their safety.

While studying meteorology, one will get familiar with the atmosphere and condition of the atmosphere at a given point of time and the different physical processes that govern given the changes in the condition of the atmosphere over a given period of time. Therefore, the study of meteorology requires continuous monitoring of the atmosphere over a wide range of locations as the atmospheric conditions may differ from place to place even at a given time. Man's curiosity to understand the changes in the atmospheric conditions from time to time necessitated the development of methodologies for predicting the condition of the atmosphere well in advance. It was realized that forecasting of atmospheric conditions has to be based on atmospheric conditions observed simultaneously across different locations in a given region or country or over the globe. In the process, scientists realized that the atmospheric conditions change over short time whereas the changes for longer periods within a year, season remains uniform for considerable number of years. Therefore, climatology was evolved as a science which deals with the average conditions of the atmosphere and the limits within which these conditions vary with respect to time and space. Thus, climatology is different than meteorology and can be divided into different areas of study.

Paleoclimatology deals with the climate over the period of earth's existence by examining records of tree rings, rocks, sediments and ice cores.

Historical climatology deals with changes in climate over longer periods for which weather records are available.

While the changes in the daily weather are sudden, the changes in the climate were believed to be very slow over hundreds and thousands of years. However, during recent times, with the rapid advancements in science and technology, human populations have started exploiting the natural resources at rapid rates thereby creating some kind of imbalances in the nature. Thus, the over exploitation of natural resources is leading to changes in the quality as well as quantum of natural resources like soil, atmosphere, water, biological diversity and so on. Clearing of forests for shifting cultivation, diversion of agricultural lands and reservoirs, mining, exploration and over utilization of petroleum products, industrialization, indiscriminate use of chemicals are contributing to deterioration of natural environment leading to changes in the climate at an alarming rate during the recent times. Therefore, major attention is now focused on understanding the climate variability and global climate change to evolve strategies not only to develop better strategies for adaptation to climate change and explore the possibilities of devising means to protect the atmosphere and environment.

Therefore, the study of meteorology and climatology which was the domain of meteorologists and climatologists is attracting the attention of all disciplines of natural, physical, biological, agricultural sciences and different branches of engineering and technology. Some elementary knowledge of physics and mathematics is considered as prerequisite for gaining an insight into the basics of meteorology and climatology.

The subsequent text is conceived and structured to enable the readers to easily understand and appreciate the basic concepts in meteorology and climatology for students of different disciplines and by no means to cater the needs of those who would like to specialize in these subjects.

2

The Atmosphere

Air is one of the primary things that make life on earth possible. The earth's atmosphere or air is made up of variety of gases and other particles. The dynamic layer surrounding the earth's surface containing various mixture of gases, moisture and solid particles (aerosols) is called atmosphere. The atmosphere is generally defined as the gaseous envelop surrounding the earth.

The formation of the earth was believed to be about five billion years ago. In the first 500 million years, a thick atmosphere emerged from the vapour and gases that were expelled from the interior of the earth. Hydrogen (H_2), water vapour, methane and carbon dioxide were some of the gases believed to be initially formed in the atmosphere due to out gassing from the interior of the earth. About 3.5 billion years ago, the atmosphere consisted of carbon dioxide (CO_2), Carbon monoxide (CO), water vapour (H_2O), nitrogen (N_2) and hydrogen present in atmospheric layers like troposphere, stratosphere, mesosphere, thermosphere (Fig. 2.1). The condensation of water vapour into water led to the formation of oceans in which sedimentation occurred and the hydrosphere was created. In the initial and ancient environment, oxygen was not present. Anaerobic is technical word which literally means without air (where air is generally used to mean oxygen and aerobic indicates the presence of oxygen). Evidence of such an anaerobic reducing atmosphere was hidden in early rock formations such as iron and uranium in their reduced states.

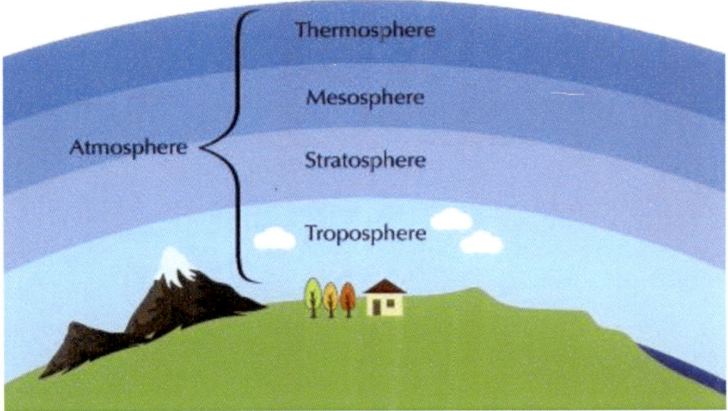

Fig. 2.1: Atmospheric Layers
https://eschooltoday.com/learn/the-atmosphere/

Elements in these states are not found in the rocks of mid-Precambrian and younger ages less than 3 billion years old. Nearly four thousand million years passed after the formation of earth before the first animals left their traces. The period prior to the observation of first animal traces is referred to as Precambrian age. About one billion years ago, early aquatic organisms called blue green algae started utilizing the energy received from the sun to split molecules of water and carbon dioxide and recombining into organic compounds and molecular oxygen. This solar energy conversion process is known as photosynthesis. Some of the photosynthetically created oxygen combined with organic carbon to reproduce carbon dioxide molecules. The remaining oxygen accumulated in the atmosphere. As oxygen in the atmosphere increased, carbon dioxide decreased. In the top layers of the atmosphere, some oxygen molecules absorbed energy from the sun's ultraviolet rays and split to form oxygen atoms. These oxygen atoms by combining with remaining oxygen molecules led to formation of ozone (O_3) molecules.

The ozone was capable of absorbing ultraviolet radiations coming from the sun, which are harmful to the life on earth. Thus, the thin layer of the ozone that surrounds the earth acts like a shield for protecting life on earth. The amount of ozone adequate to shield earth from biologically harmful ultraviolet radiation is believed to have been in existence about 600 million years back. At that point of time, the oxygen level was approximately 10 percent of the present atmospheric concentration. Earlier to this period, life and living organisms were restricted to the oceans. The development of ozone layer facilitated organisms to develop and live on land. Thus, oxygen played a significant role evolution of life on earth, and allowed life as we witness at present.

2.1 Composition of the Atmosphere

The atmosphere consists of air which is a mixture of several gases in addition to some fine solid particles suspended in the atmosphere which are known as aerosols (Fig. 2.2). So, the different gases present in the atmosphere have been discussed below:

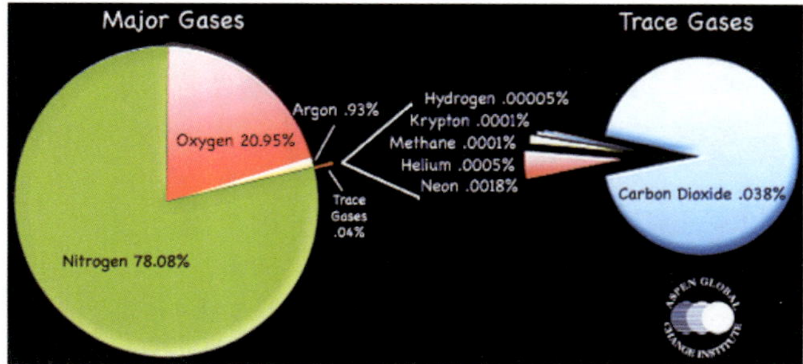

Fig. 2.2: The Composition of Atmosphere
http://apes2k15-earthsystemsandresources.weebly.com/composition-and-structure.html

Nitrogen: Nitrogen constitutes about 78% in the atmosphere. Living organisms require oxygen to make proteins. Nitrogen cannot be used directly from the air. The supply of oxygen to living organisms can be explained with the help of what is known as nitrogen cycle.

Oxygen: Oxygen constitutes about 21 percent in the air. The oxygen is required by the living organisms for respiration. Oxygen is also required for burning and combustion.

Argon: Argon constitutes about 0.9 percent of the air. Argon gas is used for filling electric bulbs.

Carbon-dioxide: The carbon dioxide constitutes 0.038 percent of the air. Plants used carbon dioxide for photosynthetic activity and release oxygen. Increased use of fossil fuels and burning of wood will increase the carbon dioxide content in the atmosphere and reduce the availability of oxygen.

Water vapour: The water vapour content varies between 0.0 to 4.0 per cent in the air. The water vapour is essential for living processes and prevents loss of heat from the earth.

Trace gases: Trace gases include neon, helium, krypton and Xenon and these are present in the air in small quantities.

In addition, atmosphere contains traces of sulphur and nitrogen compounds (sulphur dioxide, ammonia, nitrogen monoxide, nitrogen dioxide, ozone, organic halogen compounds. The atmosphere also contains solid and liquid particles of different nature as air borne particles, dust and aerosols.

The water exists in the form of liquid, vapour and ice up to about 4 per cent and most of it is generally found up to 3000 meters height. Water plays an important role in the atmosphere as transformation of water into different forms causes energy transfer in the atmosphere. The water vapour absorbs infrared radiation and regulates warming of the atmosphere.

2.2 Vertical Structure of the Atmosphere

The earth's atmosphere is divided into different layers (Fig. 2.3) based on the variation of temperature with increase in its height. The characteristics of these layers are as follows:

Troposphere: The troposphere stands at the ground level of sea surface and extends upto a height of 8 to 18 km. The important features of the troposphere are as follows:

i) The troposphere contains nearly 80 percent of the mass of the atmosphere and all most all the water vapour present in the air.

ii) The temperature decreases with increase in height of the troposphere at the rate of 6.5°C/km. The rate of decrease of temperature with increase in height is called lapse rate

iii) The height of troposphere is maximum of 18 km near the equator and minimum of about 8 km near the poles.

iv) All the major weather phenomena in the atmosphere take place in the troposphere.

v) The upper boundary of the troposphere is called tropopause.

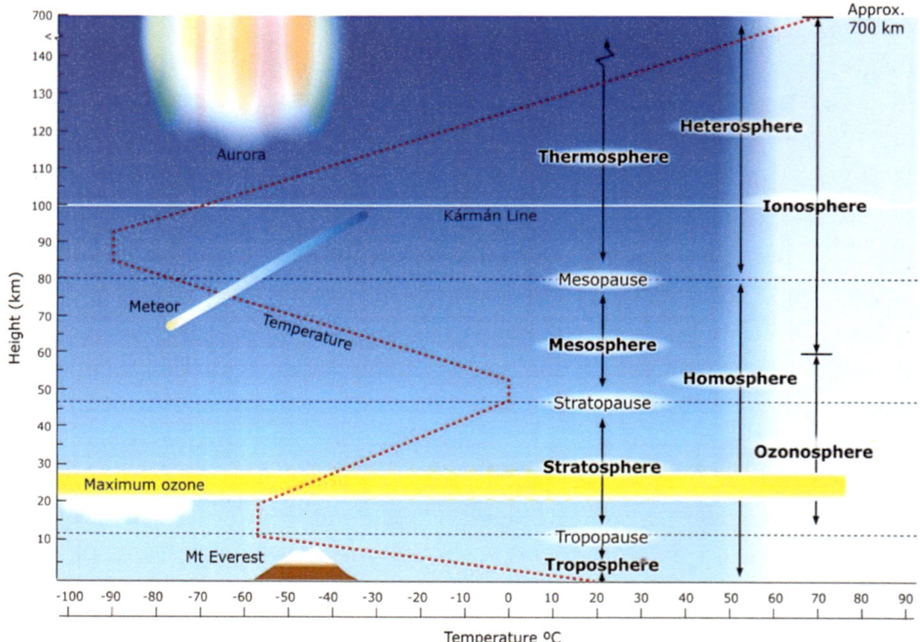

Fig. 2.3: Vertical Structure of the Atmosphere
https://www.sciencelearn.org.nz/images/240-vertical-structure-of-the-atmosphere

Stratosphere: The atmosphere above the troposphere extending up to a height of about 50 km is called stratosphere. The important features of the troposphere are as follows:

i) The ozone present in the atmosphere is concentrated in the stratosphere and the ozone content will be maximum at a height of about 30 km. As already stated, ozone absorbs ultraviolet radiation coming from the sun and protects life on earth.

ii) The stratosphere can be considered in two parts. In the lower stratosphere which exists up to about 20 km height the temperature remains constant and is about - 55°C to - 60°C. In the upper stratosphere, the temperature increases up to 0°C due to absorption of solar radiation.

iii) As the temperature increases with height in the upper stratosphere, it is called temperature inversion. The vertical movement of air is not possible in the upper stratosphere due to temperature inversion.

iv) About 99 percent of the mass of the atmosphere is concentrated within the 30km height of the atmosphere.

v) The upper boundary of the stratosphere is called stratopause.

Mesosphere: Mesosphere is the layer of atmosphere above the stratosphere extending from 50 to 85km above the ground. The important features of the mesosphere are as follows:

i) In mesosphere, the temperature decreases with increase in height. The temperature in the upper mesosphere will be about 100°C.

ii) Mesosphere is the layer in which most the meteors burn up when they enter the atmosphere as a result of collision with some of the gas particles present in this layer.

iii) The upper boundary of the mesosphere is called mesopause.

Thermosphere: The atmosphere above the mesosphere is called thermosphere. The actual temperature in the thermosphere can be as high as 200°C. It is also hot because ultraviolet radiation is turned into heat. This layer contains:

i) Ionosphere: It is the lower part of the thermosphere and extends from about 80 to 550 km above the ground. The gas particles absorb ultra-violet and x- ray radiation from the sun. The particles of gas become electrically charged ions. Radio waves transmitted from the earth are bounced back and gets reflected towards the earth. The ionosphere is thus useful in propagation of radio waves.

ii) Exosphere: It is the upper most part of the thermosphere and extends from 550km to thousands of kilometers above the ground. Air is very thin here.

Magnetosphere: The Space around the earth that extends beyond atmosphere is called magnetosphere. The earth's magnetic field operates here and begins at about 1000 km height above the ground. It is made of positively charged protons and negatively charged electrons.

2.3 Functions of the Atmosphere

Continuous monitoring of the atmospheric conditions all over the world is very important as the atmosphere -

i) Protects life on earth by absorbing all harmful radiations coming from the sun.
ii) Permits transmission of sun light in the form of solar energy to the oceans and continents.
iii) Protects the rapid cooling of earth during nights and heating during the daytime.
iv) Transfers heat energy from the equatorial regions to middle and higher latitudes
v) Transports water vapour through dynamic processes and plays important role a in the hydrologic cycle.
vi) Acts as a huge reservoir for nitrogen which is important for plants.
vii) Provides the necessary carbon dioxide, which is a basic input required for photosynthesis in plants to generate biomass.
viii) Serves as a medium for transport of pollens, insects and seed spores.

The important role played by the atmosphere is depicted in below Figure 2.4 for sustaining life on earth. The atmospheric conditions in regulating the human activities in all sphere's life, people started paying attention to:

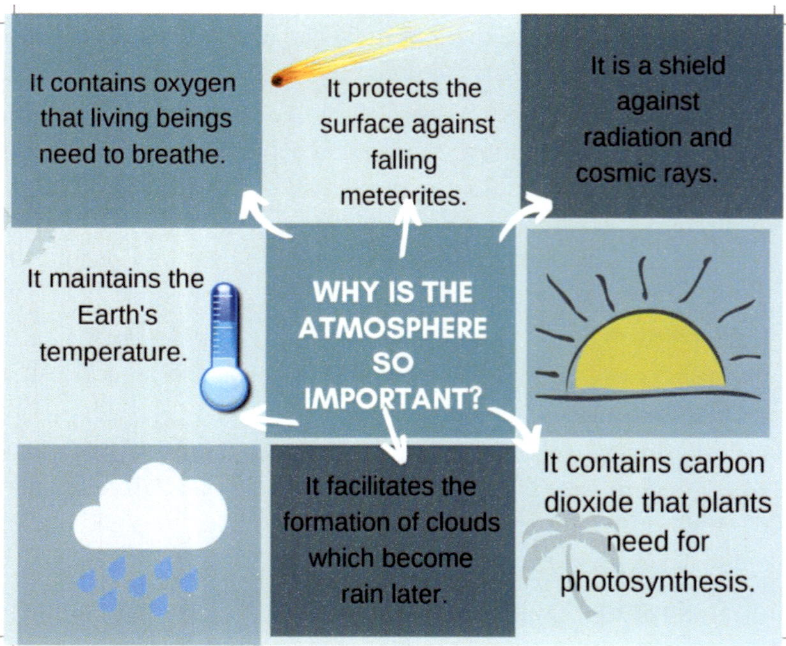

Fig.2.4: Significance of Atmosphere
https://recursos.edu.xunta.gal/sites/default/files/recurso/1605879478/ importance_of_the_atmosphere.html

i) Describe the condition of the atmosphere at a given point of time over a given place, area and region.
ii) Monitor the atmospheric conditions continuously from time to time
iii) Understand the various processes that take place in the atmosphere and predict the likely changes in the atmospheric conditions well in advance
iv) Quantify the changes in atmospheric conditions with respect to time, may be during a day, month, season or year
v) The necessity of preserving the quality of the atmosphere in terms of its constituents and structure.

2.4 Weather and Climate

In order to provide clear understanding, people are used to describe anything using a number of characteristics. If we are to describe a person whom the other person has not seen, we choose to describe in terms of height, complexion, size, colour of the hair, facial features and so on. Similarly, people started explaining the atmospheric conditions depending

i) Temperature of the air
ii) Moisture content of the air
iii) Clearness of the sky
iv) Cloudiness and movement of the air in terms of its direction and speed
v) Occurrence of rain, show, dew, frost etc. i.e., the form in which water received
vi) on earth from the atmosphere
vii) Atmospheric pressure
viii) Transparency of the atmosphere and its visibility etc.

The instantaneous condition of the atmosphere at a given point of time over a given place or area is called weather and the parameters like temperature, humidity wind speed and direction, cloudiness, the nature and type of clouds, atmospheric pressure, visibility etc. are called weather parameters.

In the early beginning, people used to talk about weather as an occasional description of words. The invention of thermometer and its regular use for observing air temperature was the first step towards quantification of weather. Galileo invented the thermometer in 1757. It became an actual physical instrument in 1780, when de Luc identified mercury as a thermometric substance. In India, rain gauges were in use as early as fourth century BC. Elements which cannot be observed using instruments are estimated with the help of arbitrary scales. When these observations are chronologically arranged, the combination of observations have become the basis for computing the averages as well as their variability with respect to time and space and these efforts have led to describe climate and development of climatology.

So, the climate is defined as the average condition of weather at a given place or region during a particular period may be a week or month or season or year.

The combination of all parameters to describe the condition of the atmosphere at a given place or area at a given point of time makes the weather. The average conditions of the weather during a specified period over a given place or region make the climate.

Thus, the weather and climate can be distinguished from each other as follows:

S. No.	Weather	Climate
1.	It is the instantaneous condition of the atmosphere at a given place at a given time	It is the average condition of the weather at a given place.
2.	The weather is described using a combination of observations made on the condition of the atmosphere at a given point of time	The climate is described based on averages of a combination of weather parameters during specified period of time period.
3.	The changes in weather refer to a particular time	The changes in the climatic conditions refer to a longer specified period of time over a larger area
4.	Weather of two places with identical values of weather parameters may be the same at both the places.	The climate of the two places having the same averages may not be the same as the distribution of the parameters during a given time scale may be different
5.	Weather is described based on observations made with instruments	Climate is described based on statistical parameters of the weather observations
6.	The changes in weather take place suddenly	The changes in the climate do not occur over short periods of time.
7.	The prediction of weather is very important in carrying out day to day activities.	Monitoring of climate is very important for long term planning purposes.

2.5 Factors Affecting Temperature at a Place

The factors affecting temperature at a place are

i) Latitude of the place
ii) Altitude of the place
iii) Distance from the sea
iv) Air mass circulation
v) The presence of warm and cold ocean currents
vi) Local aspects

Latitude is the most important factor as the amount of solar radiations reaching the earth's surface depend upon latitude. The seasonal changes in incoming radiation, length of the day vary with latitude there by affecting temperature. The past, present and future information on weather and climate is considered to be very valuable and important. Some basic knowledge on the processes that take place in the atmosphere is required to understand and appreciate the conditions that lead to changes in weather and climate in a given area/ region.

3

Physical Processes in the Atmosphere

Three fundamental processes that are responsible for inducing changes in weather are transfer of heat, energy and mass. The Sun, whose inter-surface temperature is estimated to be around 6000°C emits heat energy in the form of solar radiation which is received at the earth's surface after it is partly depleted in the atmosphere. Therefore, it is essential to understand the processes that govern the transfer of heat in the earth-atmosphere system.

The transfer of heat is governed by three different processes namely conduction, convection and radiation (Fig. 3.1)

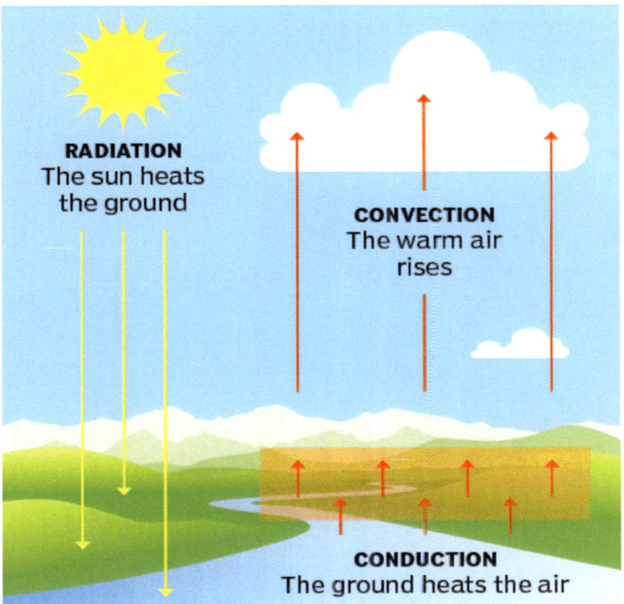

Fig. 3.1: Transfer of Heat in the Atmosphere
https://www.techtarget.com/whatis/definition/heat

3.1 Conduction

The transfer of heat from one point to another point within the material medium without the actual movement of the particles in the medium is called conduction i.e. the heat energy transfer is conducted by the particles in the medium from hot end to cold end. If a metal rod is heated at one end, the other end of rod becomes hot gradually due to transfer of heat by conduction. In the earth atmosphere system, conduction of heat is observed at the ground surface only. The amount of heat energy transferred due to conduction (Q) is

i) Directly proportional to the difference in temperature between the two points
ii) Directly proportional to the area of cross-section of the material medium through which conduction takes place
iii) Inversely proportional to the distance between the two points of the medium
iv) Directly proportional to the time

Therefore, the amount of heat transferred due to conduction can be expressed as

$$Q \propto T_2 - T_1$$

$$\propto A$$

$$\propto 1/d$$

$$\propto t$$

Where T_2 and T_1 are the temperatures at the hot end and the cold ends of the medium.

A is the area of cross-section of the medium

d is the distance between the two points of the medium and

t is the time in seconds

$$\therefore Q \propto \frac{A(T_2 - T_1)t}{d}$$

Or

$$\therefore Q = KA\,\frac{(T_2 - T_1)t}{d}$$

Where K is a constant known as thermal conductivity of the medium. The value of K is different for different materials. The materials with higher values of thermal conductivity are called good conductors and the materials which are having lower values of thermal conductivity are called bad conductors.

The thermal conductivity (K) can be expressed as

$K = Qd/A\,(T_2 - T_1)t$

The thermal conductivity of a material is thus indicates the amount of heat that is transferred due to conduction across unit area in unit time when the temperature difference is 1°C between two points separated by unit distance. The ratio of temperature difference between the two points (T_2-T_1) separated by a distance 'd' is also called as temperature gradient. Therefore, the thermal conductivity of a substance is defined as the amount of heat that can be transferred across unit area of cross section of the material in unit time when the temperature gradient is unity.

During day time when the earth's surface absorbs some of the heat energy received from the sun, the surface temperature of the earth will be more than the temperature in the lower layers immediately below the earth's surface. Therefore, there will be flow of heat due to conduction from the upper layer of the earth's surface to the lower layers. During night time, the lower layers of the earth may be warmer than the upper layer of the earth's surface and the conduction of heat may be upwards from lower layer to the upper layer. The net amount of heat energy transferred across unit area between the upper and lower region of the earth surface during the day and night together is called soil heat flux. The conduction of heat is generally observed in solids.

3.2 Convection

The transfer of heat from one point to another point due to the actual movement of the particles in the medium is called convection. When water kept in a vessel is heated at the bottom, the lower layers of water in the vessel get heated and become lighter than the denser cold layer of water at the top. Therefore, the lower layer of water gets displaced by the upper layer of the water. Thus convection of heat is generally observed in liquids and gases only. Convection of heat takes place in the atmosphere due to mixing of hot and cold air as a result of difference in pressure or gravitational force or mechanical agitation. Similarly convection of heat also takes place in water bodies like oceans, rivers etc.

3.3 Radiation

The transfer of heat from one point to another point without any material medium between the two points is called radiation. In case of radiation, the transfer of heat energy takes place in the form of electromagnetic waves with the speed of light.

The Sun is the major source of energy supply of about 99.9 per cent of total energy available at the earth's surface. The heat energy received from the Sun is called solar radiation. The nature and characteristics of thermal radiation depends upon the various properties of the surface from which it is emanated including its temperature.

In order to understand the concepts of radiation, it is necessary for us to know some of the terms used to describe the process of radiation.

3.4 Black Body

A blackbody is an ideal physical body that absorbs all incident electromagnetic radiation. As the black body can absorb all the radiation of all wave lengths, it can also emit radiation of all wave lengths. Under ideal conditions, a black body that can radiate as much heat as it absorbs at all wave lengths is called a perfect black body.

3.5 Emissivity

The emissivity of a material is the relative ability of its surface to emit energy by radiation. It is the ratio of energy radiated by a particular material to the energy radiated by a perfect black body at that temperature. The emissivity of a perfect black body is equal to 1.

3.6 Reflectivity

Reflectivity is defined as the ratio of the amount of radiation reflected by a body to the amount of radiation incident upon it.

3.7 Transmissivity

Transmissivity is defined as the ratio of heat energy transmitted to the amount of heat energy incident upon the medium.

3.8 Absorptivity

The absorptivity is defined as the ratio of amount of radiation absorbed by a body to the radiation incident upon it. If a body can absorb all the radiation incident upon it, its absorptivity is equal to 1.

3.9 Radiative Flux

Radiative flux or radiation flux is the amount of power radiated through given area in the form of photons or other elementary particles. It is measured in Watts/m^2.

3.10 Wave Length of Radiation

The wavelength is defined as the distance over which the shape of the wave repeats (Fig. 3.2 a: Long wavelength b: Short wavelength). It is also stated as the distance (measured in the direction of propagation) between two points in the same phase in consecutive cycles of a wave.

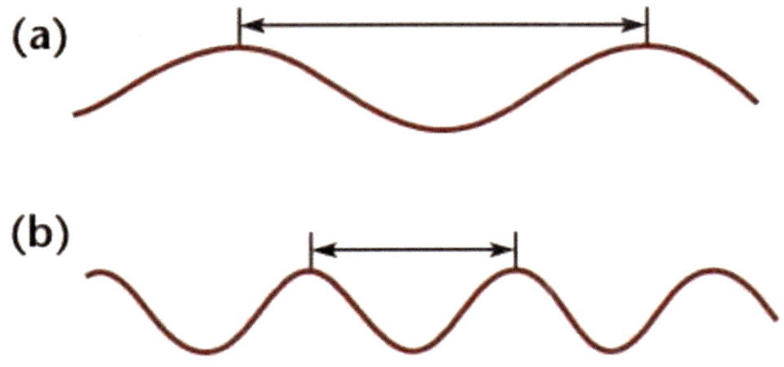

Fig. 3.2: Long and Short Wavelength
https://www.chem.fsu.edu/chemlab/chm1045/estructure.html

3.11 Frequency of Radiation

The solar energy emitted by the sun travels through space as electromagnetic radiation exhibiting wave like behavior. The electromagnetic radiation is classified according to the frequency of its wave. As the frequency increases, the wave length of radiation decreases and vice versa (Fig.3.3). The number of waves that pass a fixed point per unit time i.e., the number of cycles or vibrations undergone in unit time is called frequency. The ratio of velocity to the wavelength of the radiation gives the frequency of radiation.

Frequency of radiation =Velocity/Wavelength

The visible spectrum of radiation from 400 nm to 700 nm is also given in Figure 3.3

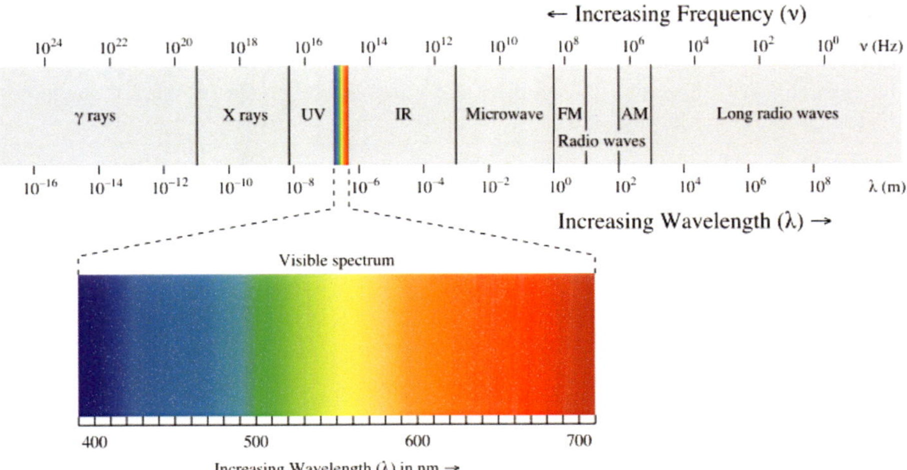

Fig. 3.3: Composition of Solar Spectrum
https://www.researchgate.net/publication/287807359 Infrared and Skin Friend or Foe

In order of increase in frequency and decreasing wave length, the electromagnetic radiation is classified as γ rays, X-rays, Ultraviolet, Visible, Infrared microwave and radio waves.

The Hertz (symbol Hz) is the SI unit of frequency and is equivalent to one cycle per second. Commonly used multiples of hertz are as follows:

1 Kilo hertz (KHz) = 10^3Hz

1 Mega hertz (MHz) = 10^6Hz

1 Giga hertz (GHz) = 10^9Hz

1 Tera hertz (THz) = 10^{12}Hz

3.12 Units of Wavelength

The wavelength of electromagnetic radiation can be expressed using the following units. However, nanometer is used as standard SI unit for expression of wavelengths.

1 meter (m) =100cm = 1m

1 centimeter (cm) = 100mm = 10^{-2}m

1 millimeter (mm) = 1000microns = 10^{-3}m

1micron (m) = 1000nanometers = 10^{-6}m

1 nanometer (nm) = 1000picometers = 10^{-9}m

A nanometer (10^{-9}m) is most common unit used for characterizing the wavelength of visible light. Using the unit, the visible portion of electromagnetic spectrum is located between 380nm – 750nm, Orange (592-620nm), Yellow (578-592nm), Green (500-578nm), Blue (464-500nm), Indigo (444-464nm) and Violet (380-435 nm0.

The electromagnetic spectrum is classified into different regions based on wavelength and frequencies as given in Table.

Region	**Frequency**	**Wavelength(m)**
Radio waves	$<10^9$	0.3
Micro waves	10^9- 3×10^{11}	0.001-0.3
Infrared	3×10^{11}-3.9×10^{14}	7.6×10^{-7}-0.3
Visible	3.9×10^{14}-7.9×10^{12}	3.8´10^{-7}-7.6×10-7
Ultraviolet	7.9 10^{14}-3.4×10^{16}	8 $\times10^{-9}$-3.8×10-7
X-rays	3.4×10^{16}-5×10^{19}	6×10^{-12}-8×10^{-9}
Gamma rays	$>5\times10^{19}$	$<6\times10^{-12}$

Thus the electromagnetic spectrum encompass a continuous range of frequencies or wavelengths of electromagnetic radiation, ranging from long wavelength radio waves to shortest wavelength gamma rays.

3.13 Laws of Thermal Radiation

The properties of thermal radiation can be explained and quantified in accordance with some rules which are generally referred to as laws of thermal radiation as explained below.

3.14 Stefan-Boltzman's Law

The Stefan-Boltzman's law is useful to estimate the total radiation emitted by a black body. The law states that energy radiated by a black body radiation per unit area (Q) is proportional to the fourth power of the absolute temperature (T) of the radiating body.

$\therefore Q \propto T^4$Or

$Q = \sigma T^4$

Where s is a constant known as Stefan-Boltzman's constant and 5,6703x10^{-8} watt/m^2/K^4

If qtemperature of the body in °C, then, the absolute temperature of the body (T) is considered as (273+q)°Kelvin.

If the body emitting radiation is not a black body, the total radiation emitted by the body

Q=EσT4

Where E= emissivity of the body emitting radiation

3.15 Planck's Law

The primary law governing black body radiation is the Planck's Law of radiation (Fig 3.4).

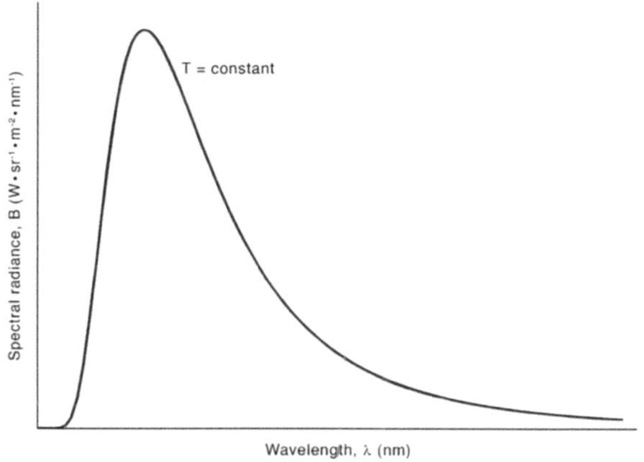

Fig. 3.4: Planck's Law
https://www.sciencefacts.net/plancks-law.html

It governs the intensity of radiation emitted by unit surface area from the black body as a function of wavelength for a fixed temperature and in simplest form, it is written as

E= (lT)= hc/λ

Where E (λ, T) = Intensity of radiation at a particular wavelength of the radiating body of temperature, T degrees absolute.

c = velocity of light ($3x10^{10}$ cm/sec)

λ = wavelength of radiation and

h = constant called Planck's constant ($6.625x10^{-27}$ergs/sec)

3.16 Wien's Law

The Wien's law establishes the relationship between the temperature of the radiating surface (T°K) and the wavelength (λ_{max}) at which the maximum intensity of radiation is emitted and expressed as

λ_{max}= 0.29/T

The Wien's law is very important as it explains that bodies at higher temperature emits high intensity radiation at short wavelengths and bodies at lower temperature will emit high intensity radiation at longer wavelengths. Thus when we compare the sun and earth, the sun emits short wave radiation and the earth whose surface temperature is generally about 30°C or 303°K emits long wave radiations.

3.17 Kirchoff's Law

Kirchoff's law deals with the nature of the material with regards for its capacity to absorb or emit radiation and it states that the absorptivity of a material for a radiation of specific wavelength is equal to its emissivity for the same wavelength at same temperature.

For a perfect black body

Emissivity = Absorptivity = 1

For a non black body

Emissivity = Absorptivity = <1

It is important to note that each law of thermal radiation has specific use. If we know the temperature of the radiating body,

-Stefan's Boltzmans is used for calculation of energy radiated by the body

-Plancks law can be used for calculation of radiation emitted at different wavelengths

-Wien's Law can be used to find the wavelength of maximum intensity radiation and

-Kirchoff's Law can be used to find the emissivity / absorptivity of the radiating body

3.18 Solar Constant

The solar energy reaching unit area at outer edge of the atmosphere exposed perpendicularly to the rays of the Sun at the average distance between the Sun and Earth is known as solar constant.

Its value is 1.4 kW/m^2 or 1.96 cal / cm^2/min.

It is also known as solar flux density.

3.19 Factors Affecting Solar Radiation

The amount of solar radiation received at a particular place and time depends upon the following factors:

i) Distance from the sun
ii) Length of the day (duration of sun light)
iii) Inclination of the sun's rays
iv) Transparency of the atmosphere

i) Distance from the sun: The planets in our solar system orbit the sun. The orbits of some planets are almost circles, but others are not circles. Some orbits are shaped like ovals or stretched out circles called ellipses. If the planets orbit is a circle, the sun will be at the centre of the circle. If the orbit is an ellipse the sun is at a point called Focus, of the ellipse. Since it is not at the centre of an elliptic orbit, the earth moves closer towards and farther away from the sun in its orbit. The place when the earth is close to the sun is called perihelion and the place when the earth is farther away from the sun is called Aphelion. When the earth is at perihelion, it is about 147 million Km from the sun. When the earth is at aphelion, the distance between the sun and earth is about 152 million Km. The earth revolves round the sun in its elliptical orbit once in a year i.e., 365 days.

Since the sun is a sphere, heat is radiated perpendicular to its surface in all directions. As the sun's rays are divergent from its surface, the amount of heat that passes through unit area decreases rapidly with the increasing distance from the sun. The flux of heat passing through a unit area is inversely proportional to the square of the distance it travels through the space.

ii) Length of the day: The earth rotates around its own vertical axis and takes 24 hours to complete one rotation. Therefore, only half the portion of the earth gets exposed to the radiation from the sun. The part that faces the sun has day and the other part not facing the Sun has night. The duration between Sun rise to Sun set i.e., the duration for which the earth's surface is exposed to sun is called length of the day. Ordinarily, one might expect that the length of the day and night are to be equal all over the earth, but, it varies. The vertical axis of the earth around which it rotates once in a day is inclined at an angle of 23½°.

When the earth is revolving round the Sun in the elliptical orbit, the north pole of the earth's axis is pointed towards the Sun on 21 June. Therefore, areas north of the equator, experiences longer day and the areas south of the equator experience shorter day length. It is called summer solstice. On 21 December, the north pole of the earth's axis is pointed directly away from the sun and therefore the areas towards north of equator experience shorter days and longer nights. It is called winter solstice.

Halfway in between the summer and winter solstices, the earth's axis points neither towards nor away from the sun. Therefore on 21st September and 21st March, the length of the day and length of the night are equal all over the earth these are called equinoxes (Fig. 3.5).

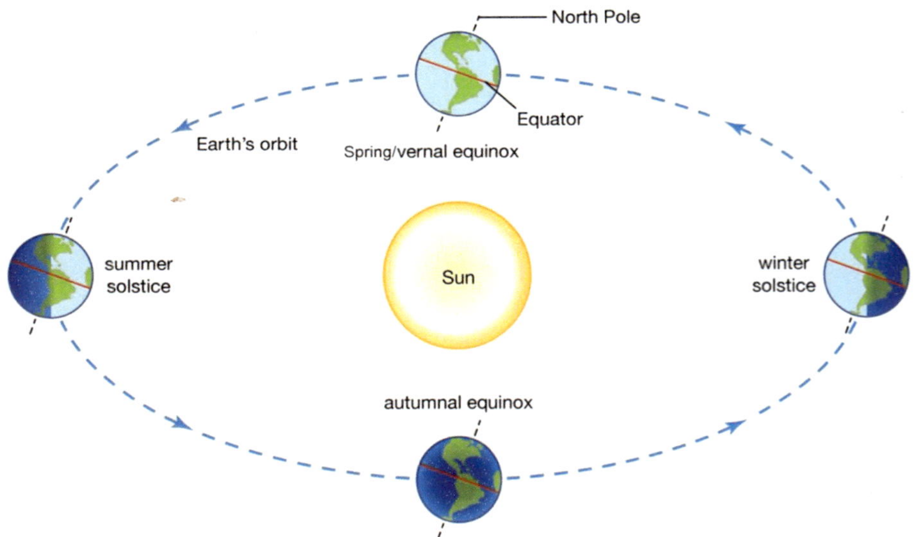

Fig. 3.5: Seasonal Configration of Earth and Sun
https://link.springer.com/chapter/10.1007/978-3-030-22403-5_1

The length of the day and night are always equal throughout the year at the equator and inequality between the length of the day and night varies more widely as we move from equator towards the poles. In the northern hemisphere the period from March 21st to June 21st during which the length of the day increases and length of the night decreases is referred to as spring season. The period from June 21 to September 23, the length of the day decreases and the length of the night increases till both are equal and this period is called summer season. During the period from September 23 to December 21st, the length of the day decreases to minimum and the length of the night increases to maximum is called Autumn season. From 21st December to March 21, the length of the night starts decreasing until both length of the day and night are equal and this period is called winter season.

However, the description of the seasons may vary some extent in the equatorial regions. From 21st March to 23rd September, the North Pole gets exposed to the sun's light all through the 24 hour period and therefore the length of the day is 24 hours. From September 21st to March 21st, the North Pole will not be facing the sun throughout the day and therefore, the length of the day is minimum and is zero.

iii) Inclination of Sun's rays: There are five major lines of latitude that are related to the inclination of Sun's rays.

Equator: Equator is the imaginary line passing around the earth's surface on the plane perpendicular to the axis joining north and south poles. It divides the earth into two hemispheres namely northern hemisphere and southern hemisphere. On 21st March and 23rd September, the Sun's rays fall perpendicular to the earth's surface at the equator during noon time.

Tropic of Cancer: The latitude of 23.5° N of equator is called Tropic of Cancer. On 21st June, the Sun's rays fall perpendicular to the earth's surface at the Tropic of Cancer during the Noon time. The inclination of Sun's rays will be 66.5° near the equator.

Tropic of Capricon: The latitude of 23.5°S is called Tropic of Capricon. The Sun's rays fall perpendicular to the earth's surface near the Tropic of Cancer during noon time on 21st December. The inclination of the Sun's rays will be 23.5° near the equator.

Arctic Circle: The latitude of 66.5°N is called Arctic Circle on 21st June, Every location north of the Arctic Circle is illuminated for the entire 24 hours and no sunlight reaches the South Pole. The inclination of sun's rays at North Pole at this time is 23.5°.

Antarctic Circle: The latitude of 66.5°S is called Antarctic Circle. On 21st December, every location towards south beyond the Antarctic Circle gets sun light all the 24 hours time and the inclination of sun's rays at the South Pole will be 23.5°.

When the sun's rays fall perpendicular to the earth's surface, it receives more radiation and as the angle of incidence changes, the amount of radiation received at the earth's surface decreases. Because of its spherical shape of the earth, incoming solar radiation is not equally distributed all over the earth. At each instance sunlight falls only on half of the earth's surface with maximum radiation coming at total noon and less in other times of the day. The total radiation receives on earth decreases from equator to poles.

iv) **Transparency of the atmosphere:** The amount of solar radiation received on unit area at the top of atmosphere in unit time when the sun's rays fall perpendicular to the surface and when the earth is at the mean distance from the sun is called Solar Constant. The approximate value of solar constant is 1.366 kW/m^2 which is equal to 1.96 cal/cm^2/minute. The solar radiation received at the outer surface of the atmosphere gets depleted to a certain extent as it passes through the atmosphere before it reaches the earth's surface (Fig. 3.6). If all the solar radiation received at the outer surface of the atmosphere reaches the earth without getting depleted while passing through the atmosphere,

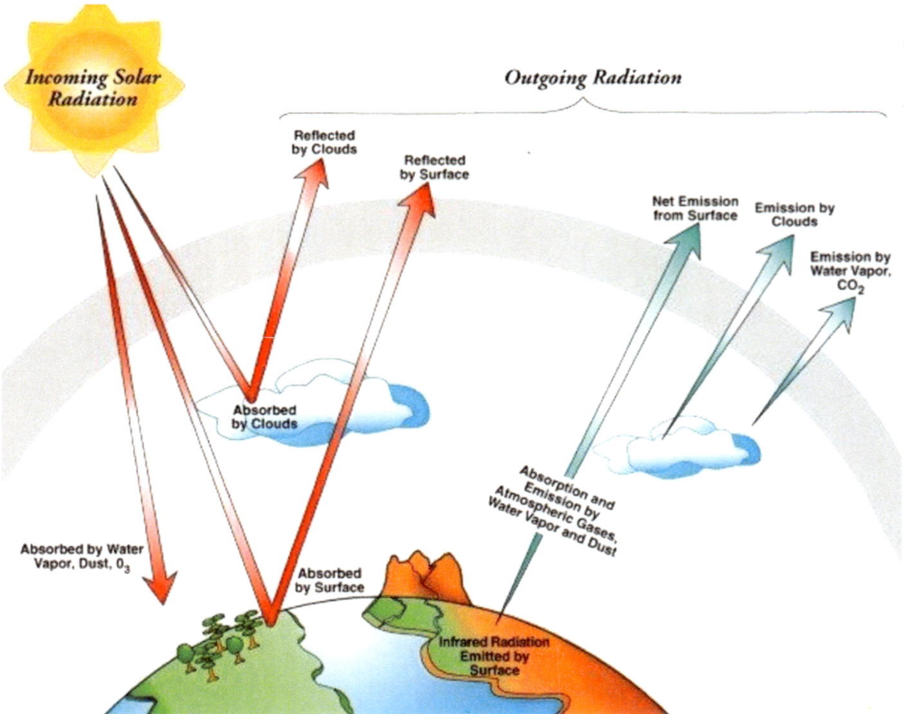

Fig. 3.6: Processes Associated with Radiation Balance
https://www.buddinggeographers.com/global-climate-vulnerability-and-resilience/

Earth would be too hot for life to exist and survive. The depletion of solar radiation is accomplished by the following four processes while passing through the atmosphere.

i) Dispersion

ii) Scattering

iii) Reflection and

iv) Absorption

i) Dispersion: As the earth's axis is inclined at angle of 23.5°, the sun's rays are received at varying angles of incidence, depending upon the position of the earth. When the sun's rays are not perpendicular to the earth's surface, the energy becomes dispersed or spread out over greater area in different directions. If the available energy reaching the atmosphere is diffused over greater area, its intensity gets reduced. Dispersion of radiation in the atmosphere is caused by the rotation of the earth and also takes place with seasons in all latitudes.

ii) Scattering: About 25 per cent of the incoming solar radiation is scattered or diffused by the atmosphere when the solar radiation passes through the air. Some of the wavelengths of solar radiation are deflected in all directions by the molecules of gases, suspended particles and water vapour. The sky appears blue as most of the radiation in the blue band of the spectrum gets scattered by the particles suspended in the atmosphere. About two thirds of the normally scattered radiation reaches the earth and it is also called as diffuse sky radiation.

iii) Reflection: When the solar radiation falls on the surface, it will get reflected and will be bounced back into space. The percentage ratio of the reflected radiation to the radiation incident upon a surface is called as albedo of the surface. All the surfaces may not have the same reflectively as stated below:

- The cloud tops will reflect about 40 to 80 per cent of the radiation and average reflectively is about 55 per cent.
- Fresh snow surfaces can reflect over 80 per cent of incoming solar radiation incident upon it.
- Land surfaces reflect from 5 per cent (when it covered by thick forest) to 30 per cent when it is dryland.
- Water surfaces reflect about 2 per cent of radiation when the sun is directly overhead and 100 percent when the sun is very low on the horizon.

iv) **Absorption:** Usually solids and liquids can absorb radiation of all wavelength whereas some gases will absorb only selected wavelengths of radiation. Some of the constituents of the atmosphere are capable of absorbing radiations at selected wavelengths. Ozone present in the layers of the stratosphere absorbs the ultraviolet radiation coming from the sun. Carbon dioxide and water vapour present in the atmosphere absorb strongly over a broad band of radiation. Clouds are by far the most important absorbers of radiation at all wavelengths. In sunlight, clouds however reflect a high percentage of incident solar radiation. The earth and the atmosphere absorb about 64 per cent of insolation. Land and water surfaces absorb about 51 per cent of the insolation.

When the sky is clear and if the water vapour content and carbon dioxide is minimum the absorption of radiation will be less and allows radiation to pass through the atmosphere undepleted.

3.20 Terrestrial Radiation

The earth's surface also emits radiation which is called as terrestrial radiation. As the temperature of the earth's surface is very much less than the sun's temperature, the earth emits mostly long wave radiation only. Most of the terrestrial radiation is absorbed by the water vapour and some of the gases present in the atmosphere and about 8 per cent is radiated back into space.

3.21 Atmospheric Radiation

The atmosphere reradiates to the outer space most of the terrestrial radiation to about 43 per cent and insolation it has absorbed (about 13 per cent).

Some of the re-radiation is emitted outward and is known as counter radiation.

3.22 Greenhouse Effect

Greenhouse gases include most diatomic gases with two different atoms (such as carbon monoxide) and all gases with three or more atoms (carbon-di-oxide, methane, nitrous oxide, ozone). These gases can absorb and emit infrared radiation. More than 99 per cent of the dry atmosphere is transparent to infra-red radiation as constituents like O_2, N_2, Argon etc. are not able to directly absorb or emit infra-red radiation.

Greenhouse is a small glass house used to grow plants especially in winter. The glass panels of the green house are transparent to radiation coming from the sun and do not allow the long wave radiation to pass through them, thereby keeping the plants warm enough to survive during winter. Similarly, when the Sun's rays travel through the atmosphere and reach the earth's surface, the earth's surface reflects a part of it and absorbs the remaining. The gases like ozone, water vapour, methane, carbon dioxide which are present in the atmosphere trap some energy coming from the sun and prevent the heat from escaping back into the space as some of these gases absorb the long wave radiation emitted by the earth. This phenomena is called as greenhouse effect as the green house is transparent to the short wave radiation and does not allow long wave radiation to escape out of it.

3.23 Global Warming

Apart from the greenhouse gases already present in the atmosphere, human activities like burning of fossil fuels, coal, natural gas, cutting and burning of trees etc are contributing to increased levels of carbon dioxide in the atmosphere. Increased use of refrigerators, air conditioners, plastics etc. contribute to increased greenhouse gases like as chlorofluorocarbons. These are harmful as even small quantities of these gases can trap large quantities of heat thereby making the earth more hotter. This phenomenon is called global warming, which is becoming a serious threat.

3.24 Solar Radiation Budget

Out of the total energy received (Fig. 3.7) at the outer surface of the earth's atmosphere, About 30 per cent of incident energy is reflected by

- Atmosphere (6%)
- Clouds (20%)
- Earth's surface including water bodies (4%)

About 70 per cent of the incident energy is absorbed by

- Land and water (51%)
- The 51% of the radiation absorbed by land water is disposed in the following ways.
- 23 per cent transferred back into atmosphere was latent heat due to evaporation of water.
- 7 per cent back into the atmosphere due to exchange of heat as sensible heat flux.
- 6 per cent radiated directly into space
- 15 per cent transferred into the atmosphere by radiation, then reradiated into space.

Of the 19% of the incoming radiation absorbed by the clouds and atmosphere.

- 16% reradiated into space
- 3% transferred to clouds where it is radiated back into space.

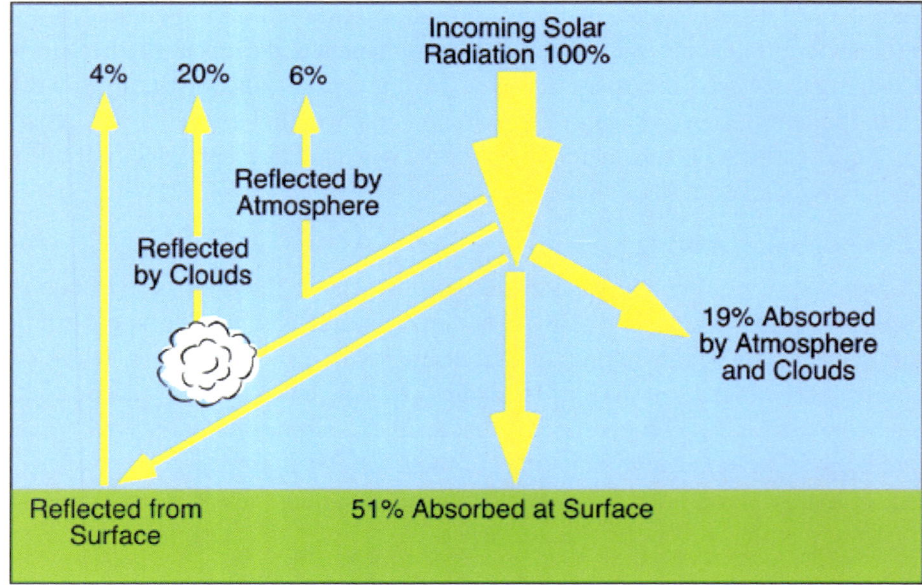

Fig. 3.7: Solar Radiation Budget
https://www.solpass.org/science6-8-new/standards/standard_6.3.html

3.25 Net Radiation

The difference between the incoming solar radiation and outgoing radiation is called net radiation.

3.26 SI Units used to Measure Thermal Radiation

The thermal radiation expressed using the following units.

i) Calories / cm^2

ii) Kilo Calories / cm^2

iii) Joules/ m^2.

The amount of heat required to raise the temperature of one gram of water by one degree centigrade is called Calorie. Thousand calories is equal to one kilo Calorie. One Joule is equal to 4.1868 joules or approximately 4.2 Joules.

4

Temperature and Heat Transfer in the Atmosphere

In meteorology, temperature is always referred to as the temperature of the air unless it is stated otherwise. There will always be exchange of heat in the atmosphere from a point at high temperature to a point at lower temperature. The transfer of heat through convection and radiation is always observed in the atmosphere whereas the conduction of heat takes place near the earth's surface.

4.1 Diffusion of Heat

Heat diffuses (spreads) at different rates through different media, the rate at which diffusion of heat takes place is called thermal diffusivity. The thermal diffusinty of of some selected materials is given below:

Material	Thermal diffusivity (cal/cm^2/sec)
Wood	0.00128
Marble	0.0120
Iron	0.2034
Aluminium	0.8418
Iron	1.7004

One can see from the above table that iron has got higher diffusivity compared to wood. Suppose one end of iron rod is kept at hot temperature, the heat spreads faster in iron rod and the other end gets warmer quicker than in the case of a wooden stick which means that the transfer of heat is quicker in bodies with higher thermal diffusivity.

4.2 Thermal Resistance

The thermal resistance is a measurement of a material's resistance to heat flow. It is reciprocal of thermal conductance. It is calculated using the formula –

R=L/K

Where R= thermal resistance

L= thickness of material

K= thermal conductivity

The concept of thermal resistance is used in electric industry.

4.3 Specific Heat of Gases

In case of gases, two types of specific heat are considered.

4.3.1 Principal Specific Heat of Gas at Constant Volume (Cv):

It is defined as quantity of heat of a gas at constant pressure absorbed or released for rise or fall in temperature of unit mass of gas by 1°C when its volume is kept constant.

4.3.2 Principal Specific Heat of gas at Constant Pressure (Cp)

The principal specific heat of a gas at constant pressure is defined as the quantity of heat absorbed or released for rise or fall in temperature of unit mass of gas by 1°C when its pressure is kept constant.

4.4 Relationship Between Specific Heat of Gas at Constant Pressure and Specific Heat of Gas at Constant Volume

The relationship between specific heat of gas at constant pressure and specific heat of gas at constant volume is given by

Cp- Cv =R

Where, R is called gas constant.

4.5 Molar Specific Heat of a Gas at Constant Volume

The molar specific heat of a gas at Constant volume is defined as the quantity of heat absorbed or released for rise or fall in temperature of one mol of gas by 1°C when its volume is kept constant.

4.6 Molar Specific Heat of A Gas at Constant Pressure

The molar specific heat of a gas at constant pressure is defined as the quantity of heat absorbed or released for rise or fall in temperature of one mol of gas by 1°C when pressure is kept constant.

The difference between the molar specific heat of gas at constant pressure and molar specific heat of gas at Constant volume is called molar gas Constant and it is equal to 8.31446262 Joules / degree K/mol. In this case degree K represents the degree absolute in Kelvin scale of temperature.

4.7 Thermodynamics

Thermodynamics explains how thermal energy is converted to or other forms of energy and how matter is affected by the process. There are four different types of thermodynamic processes.

4.8 Adiabatic Process

In adiabatic process, there is no heat transfer into or out of the system.

4.9. Isochoric Process: In Isochoric process, there is no change in volume and the system does not do work.

4.10. Isobaric Process: In Isobaric process, no change in Pressure takes place.

4.11. Isothermal Process: In Isothermal process, no change in temperature takes place.

4.12. Enthalpy: Enthalpy is the measurement of energy in a thermodynamic system. The quantity of enthalpy equals to the total heat content of a system equivalent to the sum of systems internal energy and the product of Volume and Pressure. It is expressed as

$H = E + PV$

Where H = Enthalpy

P= Pressure

V= Volume

E = Internal energy

4.12.1 Entropy

Entropy is a scientific concept that is most commonly associated with a state of disorder, randomness or uncertainty. In thermodynamics, entropy is a systems thermal energy per unit temperature which is not available for doing useful Work.

4.13 Laws of Thermodynamics

The thermodynamic processes are governed by a set of rules which are called laws of thermodynamics.

4.13.1 First Law of Thermodynamics

The first law of thermodynamics states that energy can neither be created nor destroyed, but it can be transformed from one form to another.

Mathematically, the first law of thermodynamics can be expressed as

$\Delta u = q + w$

Where

Δu= the total change in the energy of the system

q = heat exchanged between the system and surroundings

w=work done or by on the system.

The work is also equal to the product of negative external pressure of the system and change in volume.

$W=P.\Delta V$

Where

P = external pressure of the system

ΔV = Change in Volume

The internal energy of the system decreases if it does work. The internal energy of the system increases if the work is done on the system.

4.13.2 Second Law of Thermodynamics

The second law of thermodynamics states that heat energy cannot be transferred from a body at a lower temperature to a body at higher temperature without the addition of energy.

It is difficult to turn all the heat emitted by a hot body into work. The working material of a heat engine absorbs heat from a hot body, transforms a portion of it into work and returns the remainder to the cold body. No engine can transform all the heat from the source into work without wasting any heat. It is theoretically not possible to convert energy with 100 percent efficiency into work. Therefore, entropy describes how much portion of energy cannot be converted into work. The change in entropy of the system is given by

$\Delta S= Q/T$

Where

Q = the heat that transfers energy during the process i.e., the heat energy used to perform work.

T = Absolute temperature at which the process takes place.

ΔS = Change in entropy

The standard unit of entropy is Joule /Kelvin degree of temperature.

4.14 Latent Heat

Latent heat is defined as the heat or energy that is absorbed or released during phase change of a substance without change of temperature of a substance. It can be from gas to liquid or liquid to solid and vice versa.

4.14.1 Latent Heat of Vaporization

The latent heat of vaporization is defined as the amount of heat required to convert 1gm of liquid into vapour at the boiling point of the liquid.

The latent heat of vaporization of Water is 540 cal/gm.

4.14.2 Latent Heat of Fusion

Latent heat of fusion is the amount of heat required to convert 1gm of solid to liquid at its melting point.

The latent heat of fusion of ice 80 Cal/gm.

Therefore, it can be observed that there will be transfer of heat in the atmosphere as water changes its phase. For e.g. when water vapour condenses into water drops, the water vapour releases heat to the air in the atmosphere. Similarly, when water drops condense into ice crystals in the atmosphere the heat released by ice as latent heat and provides energy for several processes in the atmosphere.

4.15 Sensible Heat

Sensible heat is referred as a form of energy that is emitted into the atmosphere or absorbed. The sensible heat can be calculated by using the formula

$Q = mc\Delta T$

Where

Q = Sensible heat

m = mass of the body

C = specific heat capacity of the body

ΔT = change in temperature

4.16 Difference Between Latent Heat and Sensible Heat

Sr No.	Latent heat	Sensible heat
1.	Energy released or absorbed due to change in the phase of the substance.	Energy released o absorbed without change in phase of the substance.
2.	It is change in internal energy of the substance.	It is energy absorbed or reloaded due to change in temperature.
3.	It is a process with in the substance.	It is a process between the substance and its surroundings.
4.	In Case of latent heat, no Change in temperature of the substance.	In case of sensible heat there is change in temperature of the substance.

4.17 Potential Temperature

Potential temperature is defined as the temperature attained by air parcel of temperature T and Pressure P would have if it were expanded or compressed under adiabatic conditions to standard pressure which is usually taken as 1000 milli bars.

4.18 Virtual Temperature

Virtual temperature is the temperature that dry air would have if its density and pressure were equal to those of a given sample of moist air.

The potential temperature (θ) of a parcel at pressure P would attain if adiabatically brought to Pressure Po is given by

$\theta=T(P_o/P)^{R/cp}$

Where

T= Temperature of the air parcel

R= gas Constant

Cp= The value of R/cp is taken as 0.286 for air in meteorology.

The virtual temperature is calculated using the formula

$T_v=T(1+r_v/E)/(1+r_v)$

Where

r_v= mixing ratio

E = ratio of gas constants of air and water vapour which is about 0.622

T_v= Virtual temperature of the air

T = Temperature of the air.

4.19 Gas Laws

There are set of rules that govern the relationship between the pressure, volume and temperature of the gases. These are called gas laws.

There are four fundamental gas laws to discover the relationship of pressure, temperature volume and amount of gas.

i) **Boyle's Law**: The volume (v) of a given mass of gas inversely Proportional the pressure of the gas at a constant temperature.

   $V \alpha 1/P$ or $P\alpha 1/V$ or $PV=K_1$

   where K_1 – Constant

   P = Pressure of the gas

   V = Volume of the gas

   According to Boyle's Law the volume of a given mass of gas (V_1) and pressure (P_1) at a given temperature T will have its volume V_2 when the pressure is changed to P_2

   $P_1V_1 = P_2V_2$

   of pressure increases, its volume decreases and vice versa at a given temperature.

ii) **Charle's Law**: According to Charles Law, the volume of a given mass of gas is directly proportional to the absolute temperature of the gas at a given pressure

$P \alpha V$

Where, V = Volume of the gas

T = Absolute temperature of the gas

iii) **Gay-Lussac Law:** The pressure of a given amount of gas held at constant volume is directly proportional to the absolutetemperature of the gas.

$P \alpha T$

where P = Pressure of the gas

T= Absolute temperate of the gas.

iv) **Avagadro's Law:** According to Avogadro's law, equal volumes of different gases Contain equal number of molecules when the temperature and Pressure remain the same.

$V \alpha n$

Where, V = Volume of the gas

n = amount of gaseous substance expressed in moles.

v) **Universal Gas Equation:** Universal gas equation can be obtained by combining the Boyle's law, Charlee's law and Avogadro's law as shown below:

$V \alpha 1/P$ Boyle's, law

αT Charles law

αn Avogadro's law

or $PV \alpha n T$

or $PV = n R T$.

Where, R = Universal gas constant. 8.3144 598 J/mol/ degree A.

The transformation of energy. in the atmosphere leads to changes in atmospheric pressure, temperature, movement of air, formation of clouds, precipitation, movement of air masses, movement of cyclones etc. It is conversion of heat energy received from the Sun that contributes to all the physical processes that take place in the atmosphere. In this context, it is necessary for us to get familiar with the thermodynamics as explained in the preceding sections.

5

Atmospheric Pressure

The mass of a body is usually defined as the material content of the body whereas the weight of the body is defined as the force with which the body is attracted towards the earth due to gravity. Therefore, the weight of a body is the product of its mass and acceleration due to gravity. The air in the atmosphere, which is a mixture of several gases has certain mass and although not very dense, will acquire some weight due to gravitational force of attraction when the entire column of the atmosphere from the ground level to the top of it is considered. Pressure is defined as the force for unit area. Therefore, the atmospheric pressure is defined as the force exerted by an air column of units area of cross section extending from earth's surface to the top of atmosphere. As we go up in the atmosphere the height of the air column decreases, the air also become rarer and the atmospheric pressure decreases with increase in height.

The standard atmospheric pressures is defined as the atmospheric pressure measured at mean sea level at 45° Latitude when the temperature is 273°K (=°C).

The standard atmospheric pressure is numerically equal to the pressure exerted by a mercury column of height 76 cm at 0°C

Standard atmosphere Pressure (P) = hdg

Where h = height of the mercury column 76 cm

d = density of mercury at 0°C (13.5951 gm/cm^3)

g = acceleration due to gravity 980.665cm/sec^2

P = 76x13.5951x980.665 =1013250 dynes/sq. cm

The atmospheric pressure is usually expressed in millibars

1 milli bar = 1000 dynes/sq. cm

1 bar = 10^6 dynes/sq. cm

The other units used for pressure are as follows:

1 Pascal = 10^{-2} milli bar = 10 dynes/sq. cm

1 mill bar = 1 hacto Pascal (hpa)

1 torr = 1333.33 dynes/sq. cm

1 atmosphere = 760 torrs = 1013250 dynes/sq. cm or 1013.25 mb

The atmospheric pressure describes the tendency of air to rise or sink at a given place or time. Air tends to rise or sink depending upon its density. The density of air decreases with increase in height. But, the density of air at ground level depends upon the temperature. The atmospheric pressure decreases on the average at the rate of 34 mb for every 300 meters of height.

5.1 Reduction of Pressure to Sea Level

As the surface of the earth is not flat and the atmospheric pressure decreases with increase in height, it is difficult to compare the atmospheric pressure measured at different locations and different levels above the mean sea level. Therefore, we have to reduce the pressure to sea level i.e. we have to calculate what would have been the atmospheric pressure, if the locations are at mean sea level. The term reduction is used irrespective of the fact whether the original pressure is obtained from above sea level or below sea level.

The equation used to calculate the pressure at mean sea level (Po) is called hypsometric equation and is written as

$$Po = P\left(e\frac{gz}{RdT}\right)$$

where P = surface pressure measured at a height 'Z' above mean sea level

g = acceleration due to gravity

Rd = gas constant for dry air (2.870x102J/(Kg*K)

T = Mean layer temperature from the surface to sea level

5.2 Diurnal Variation of Atmospheric Pressure

The atmosphere pressure shows a semi-diurnal variation arising from atmospheric tides, with maximum at around 0000 and 1200 hrs local time and minimum at around 0800 and 2000 hrs (Fig. 5.1). The magnitude of semi-diurnal variation is more near the equator compared to the poles.

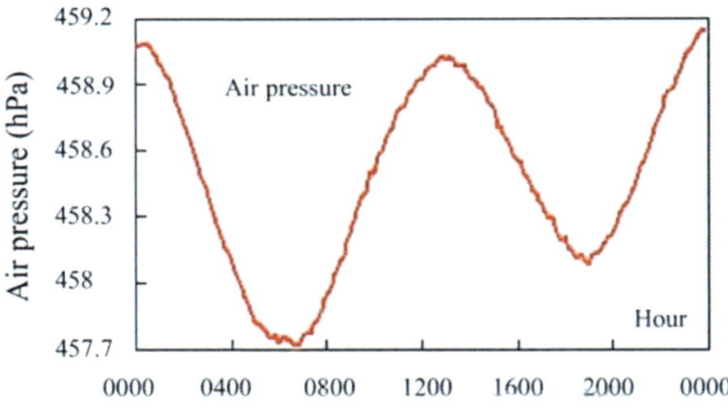

Fig. 5.1: Air Pressure Variation

https://www.researchgate.net/publication/225458262_Meteorological_features_at_6523_m_of_Mt_Qomolangma_Everest_between_1_May_and_22_July_2005

5.3 Pressure Gradient

The pressure gradient is defined as the ratio of the difference in atmospheric pressure between two points in the same horizontal plane to the distance between the two points. If the atmospheric pressure at mean sea level is same at all places in the earth surface, there will not be any pressure gradients. The air in horizontal motion is called wind and the air generally moves from a high-pressure area to a low-pressure area just as water flows from higher level to lower level. Therefore, the air movement as wind occurs due to differences in atmospheric pressure from place to place.

5.4 Atmospheric Circulation

Let us start with the assumption that earth is homogenous and stationary. Then the earth receives more heat due to solar radiation near the equator and less heat due to solar radiation near the poles. The warm air at the equator will rise vertically and the cold dense air near the poles will sink i.e, descends downwards. The cold air moves from poles towards the equator at the surface level and the warm air moves from the equator to the poles at upper level (Fig. 5.2). Therefore, the circulation of air between the equator and poles is expected to take place in a single cell.

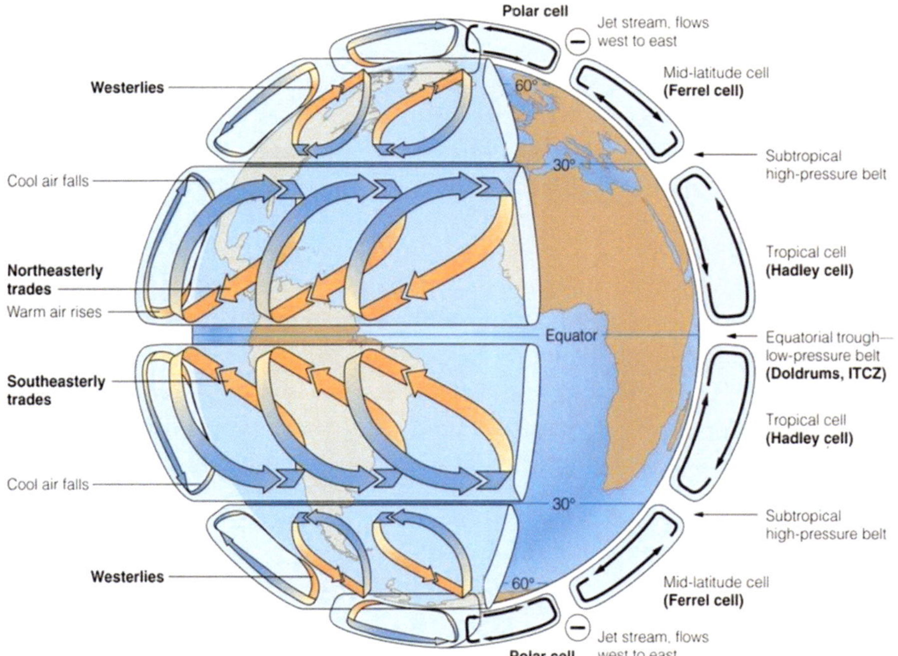

Fig. 5.2: Global Wind Systems
https://www.researchgate.net/publication/329138708_Understanding_seismic_body_waves_retrieved_from_noise_correlations_Toward_a_passive_deep_Earth_imaging/figures?lo=1

5.5 Coriolis Effect

In reality the movement of the air does not take place exactly from north to south or from south to North. Instead, the air seems to move east to west or west to east. Because of rotation of the earth around its own vertical axis, any moving material above the earth surface tends to drift sideways from its direction of motion. In the northern hemisphere, the movement is towards the right direction of motion and in the southern hemisphere, the movement will be towards the left of the direction. This phenomenon is called Coriolis effect. The force that diverts the wind direction laterally due to rotation of the earth is called Coriolis force. The Coriolis force thus diverts the wind towards the right in the northern hemisphere, and towards the left in the southern hemisphere.

The effect of Coriolis force is more at higher altitudes, as the friction force at ground level will be more significant. The Coriolis force (F_C) is given by the formula

$F_C = -2m\ (v \times ɯ)$ where

v - velocity of moving object

ɯ–Angular velocity of rotating earth 7.27×10^{-5} rad/sec

Coriolis force is observed due to rotational movement of the earth. Coriolis force is affective only when object is in motion, Coriolis force affects wind direction and not wind speed as it deflects the wind. As the wind speed increases, the deflection of wind will be more as observed in case of cyclones. The Coriolis force is zero at the equator and maximum near the poles as the angular velocity of earth increases with increase in latitude.

5.6 Geostrophic Wind

As the air moves from high pressure region to low pressure region under pressure gradient force the direction of the wind gets changed due to Coriolis force. If there is balance between the pressure gradient force and the coriolis force, the wind moves along the isobars. Isobars are the lines joining equal atmospheric pressure. The wind is called geostrophic wind. However, it is an ideal condition as the wind direction is also affected by the frictional forces near the ground.

5.7 Effect of Friction Near the Ground

Surface winds on a weather map do not flow exactly parallel to the isobars in the case of geostrophic wind. Instead, the surface winds cross the isobars at an angle between 10 to 45°. Near the earth's surface, friction due to uneven surface of the earth reduces the wind speed and reduces the coriolis force. The frictional force acts in a direction opposite to the direction of the wind.

A calm ocean surface may be smooth and offers little friction. As we go up in the atmosphere, the effect of frictional force decreases until the wind becomes geostrophic. The level at which the frictional force will not exert influence on the wind and the wind becomes geostrophic is considered as the top of the atmospheric boundary layer. Usually the boundary layer exists at a height of 1 to 2 Km above the earth's surface.

5.8 Gradient Wind

While the pressure gradient force causes air to flow from high pressure to low pressure across the isobar pattern, another force acts upon the wind to determine its direction.

The Coriolis force created by earth rotation, diverts the air towards the right with respect to its initial direction of motion even when the air is near high- or low-Pressure system. The result of the two forces is gradient wind which flows perpendicular to the pressure gradient force This means that the gradient winds flow parallel to the isobars and results in circulation of air. The gradient winds are formed 2000 ft above ground level.

5.9 Cyclostrophic Wind

The wind circulation that results from a balance between local atmospheric pressure gradient and the centrifugal force are cyclostrophic winds and or observed near the equator where there is no coriolis force (Fig.5.3).

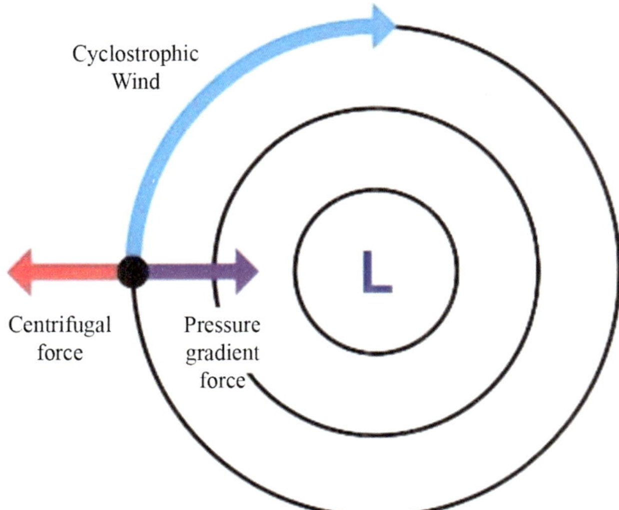

Fig. 5.3: Cyclostrophic Wind
https://www.britannica.com/science/cyclostrophic-wind

5.10 Pressure Systems

Atmospheric pressure at the earth's surface is the primary parameter which provides clue to weather. By plotting the atmospheric pressure measured at different locations on a synoptic chart, the regions with high atmospheric pressure are indicated with the letter H and the regions with low atmospheric pressure are indicated with letter 'L' (Fig. 5.4). The high-pressure areas marked 'H' indicate that the atmospheric pressure in the region is more than that of surrounding areas. Similarly, the low-pressure areas marked 'L' indicate that the atmospheric pressure in the region is less than that of surrounding areas.

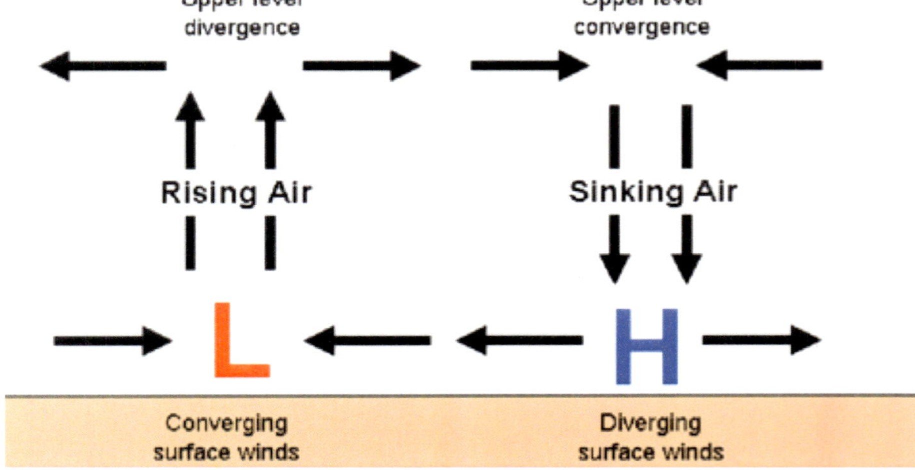

Fig. 5.4: Low Pressure Area with Convergence of Winds and High Pressure Area with Divergence of Winds
https://quizlet.com/788076711/weather-and-climate-exam-2-flash-cards/

High pressure areas will have sinking air and are generally associated with fair weather conditions. The high-pressure systems usually cover larger areas compared to low pressure systems, move slowly and will have longer atmospheric life. These high-pressure regions are also known as anticyclones (Fig. 5.5). As air descends down due to sinking, it warms up. The water droplets in the air evaporate and leads to dry weather.

The low-pressure systems are characterized by rising air. A low-pressure system develops when relatively warm and moist air rises from the earth's surface. As the air moves upwards, it gets cooled and clouds form. Therefore, cloudy conditions, windy weather and precipitation in the form of snow or rain usually occur in case of cyclones (Fig. 5.5).

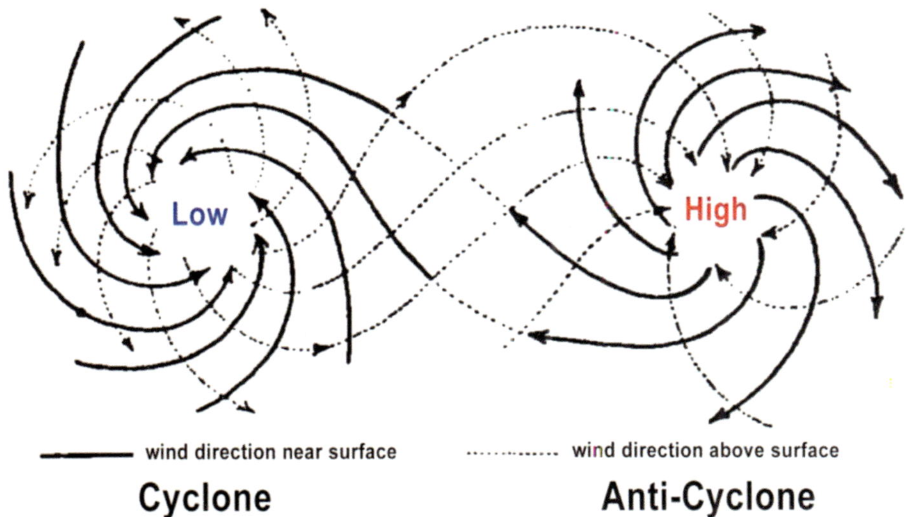

Fig. 5.5: Circulation of Winds Due to Cyclone and Anticyclone in NH
https://commons.wikimedia.org/wiki/File:Cyclone-anti-cyclone-wind-direction-air-pressure.jpg

There is a basic difference in the low-pressure systems of tropical and extra tropical regions. The terminology used for calling low pressure systems is as follows:

5.11 Cyclone: The cyclones will have circulating winds near the ground with anti-clockwise direction in the northern hemisphere and clockwise circulation of winds in the southern hemisphere. These are also called typhoons.

5.11.1 Tropical Cyclone: The cyclones formed in tropical areas are known as tropical cyclones.

5.11.2 Typhoons: Typhoon is a tropical cyclone located in the western north pacific basin (between 100°E and 180°E) in the northern hemisphere.

5.11.3 Hurricane: Hurricane is a tropical cyclone in north Atlantic eastern or central north pacific (east of 180°E) in the northern hemisphere, eastern south pacific (east of 160°E) in the southern hemisphere

5.11.4 Tornados: Tornados are associated with violent circulation of winds touching the ground and extend up to the base of a thunder cloud. Tornados are more violent and destructive with wind speeds even more than 300 miles per hour and it will have a life span if just 10 to 15 minutes. The circulation of winds will be in the shape of a funnel with narrow end on the ground. The tornados can cause heavy damage to the life and permanent structures. Tornados occur in North America more frequently.

5.12 Structure of Cyclones: The center of a cyclone is a calm area. It is called eye of the cyclone. A cyclone is a rotating mass of air in the atmosphere, 10 to 15 km above the earths surface. Around the calm and clear eye, there is a cloudy region of 150 km in size.In this region there will be high wind speeds ranging from 150 to 200 kmph and thick clouds with heavy rain. Away from this region ,the wind speed gradually decreases.

5.13 Anti cyclones: The anti-Cyclone is large wind system associated with high pressure with circulation of winds in the clockwise direction in the Northern Hemisphere and anti-clockwise direction in the Southern Hemisphere.

5.14 Trough: Trough is a low-pressure system but it is not closed isobaric system. It is a line along which the direction of wind changes abruptly. The wind blows in the anti-clockwise direction, in the Northern hemisphere and clockwise direction in the southern hemisphere.

5.15 Ridge: Ridge is a high-pressure system but not a closed isobar. It is also a line along which the direction of winds changes abruptly. In case of ridge, the winds blow in the clockwise direction in the northern hemisphere and clockwise direction in the southern hemisphere.

5.16 Col: It is a region of intersection between trough line and ridge line and a region between two high pressure areas to low pressure areas (Fig. 5.6). Near the center of the col, the pressure gradient is very weak and wind speeds are very less.

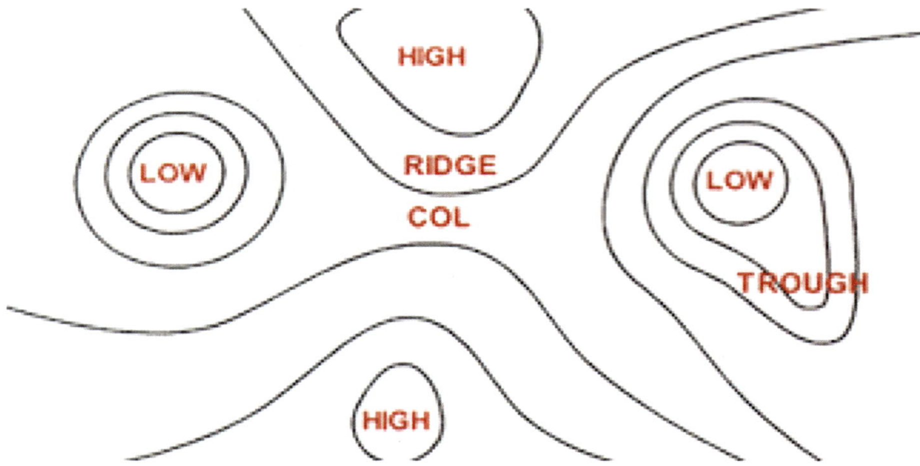

Fig. 5.6: Pressure Distribution Associated with Col
https://future-seafarer.com/col/

5.17 Blizzard: Blizzards are dangerous winter storms that consist of blowing snow and wind resulting in very low visibility. A ground blizzard occurs when strong winds blow loose snow particles are lifted from the ground. Blizzards scan stretch from 100 to even 1000 miles and will last for several hours.

5.18 Water Spout: The water spout is a whirling column of air and mist. Some of the water spouts can be as dangerous as tornados. The water spouts are classified as two types i.e., fair weather water spouts and tornadic water spouts. Tornadic water spouts are tornados that form over water or move from land to water. It is a tornado in water.

5.19 Avalanche: Avalanche is a large mass of snow, ice, earth, rock or other material in swift motion down a mountain side. The causes of an avalanche include terrain, layers of snow pack, weather, wind, heavy rains and disturbances.

The angle, orientation, shape and features of slope indicate the avalanche potential. The regions vulnerable to avalanches are characterised on the basis of:

- A steep slope of greater than 30°
- Multiple starting zones
- Little or no tree cover
- Gullies and valleys in the topography that can channel the movement of avalanche debris

5.20 Classification of Low-Pressure Systems in India

The low-pressure systems are classified on the basis of pressure deficiency when the system is on the land and based wind speeds when the system is over the sea as follows:

System	Pressure deficient (hPa)	Associated wind speed (Kmph)
Low pressure area	1.0	<32
Depression	1.0-3.0	32-50
Deep depression	3.0-4.5	51-59
Cyclonic storm	4.5-8.5	60-90
Severe cyclonic storm	8.5-15.5	90-119
Very severe cyclonic storm	15.5-65.6	119-220
Super cyclonic storm	>65.6	>220

5.21 Global Pressure Belts

There are four distinct pressure belts that prevail over the globe distributed according to latitudes (Fig. 5.7).

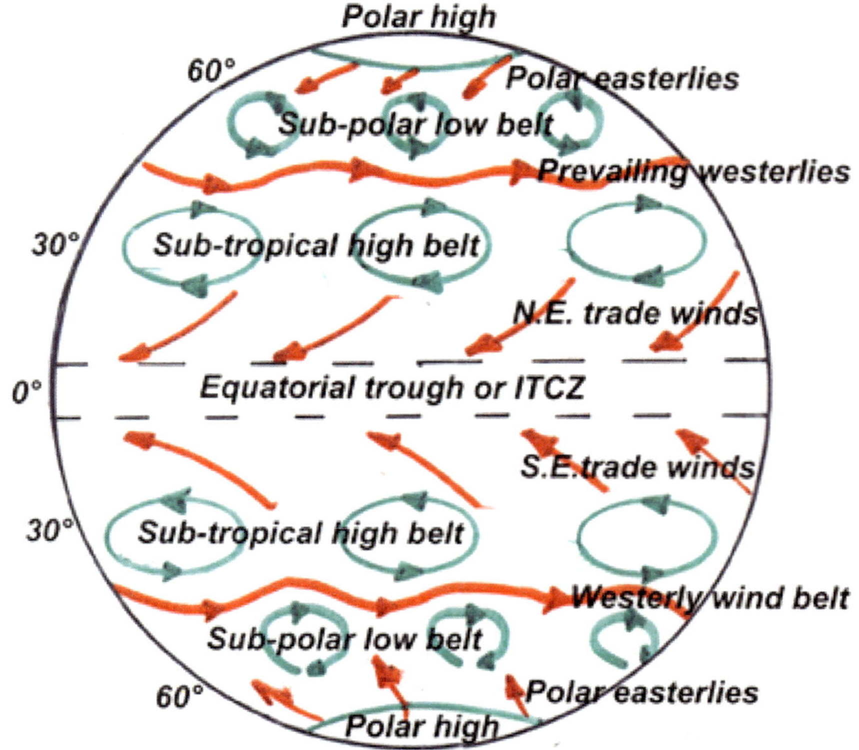

Fig. 5.7: Global Pressure Belts
https://testbook.com/ias-preparation/pressure-belts-of-the-earth

i) **Equatorial Low-Pressure trough:** The equatorial low-pressure belt lies within few degrees on both sides of equator where the atmospheric pressure is slightly less than the normal pressure of 1013mb. This area is characterized by high temperatures, more moisture content in the atmosphere and is commonly known as Doldrums. The low pressure in the area is due to more heating. The low-pressure system is shallow and influences surface wind patterns.

ii) **Sub-Tropical High-Pressure Belts:** The sub-tropical high-pressure belts are observed at about 30° latitude both in the northern and southern hemisphere. These latitudes are called "Horse Latitudes". These zones are characterized by calm and descending winds. These are developed due to dynamic process rather than thermal causes. The high pressure is due to descending air.

iii) **Sub Polar Low-Pressure Belts:** These occur at about 60° North or south latitudes and are characterized by cool and wet weather. These are caused by the meeting of cold air masses from higher latitudes and warm air masses from lower latitudes. The polar front forms in the northern hemisphere due to meeting of these air masses. As a result, low pressure cyclonic storms are causing rainfall in Europe and Pacific Northwest regions. In the southern hemisphere, severe storms develop along the fronts causing strong winds and snowfall in Antarctica.

iv) **Polar High-Pressure Belts:** These High-Pressure belts are observed in the polar regions at 90° N or S Latitudes and are characterized by cold and dry winds moving away from the poles. These are generally weak as the solar energy available at the poles is comparatively very less.

5.22 Global Wind Patterns: The global wind pattern is popularly known as general circulation and the surface winds in both the hemispheres are divided into three wind belts (Fig. 5.8).

i) **Polar Easterlies:** The polar easterlies are the dry and cold winds that blow from high pressure areas of the Polar Highs at the north and south poles towards low pressure areas in the sub-polar regions. In Polar regions, the cold air descends and causes movement of air towards sub-polar regions and these winds blow from east to west under the influence of Coriolis force.

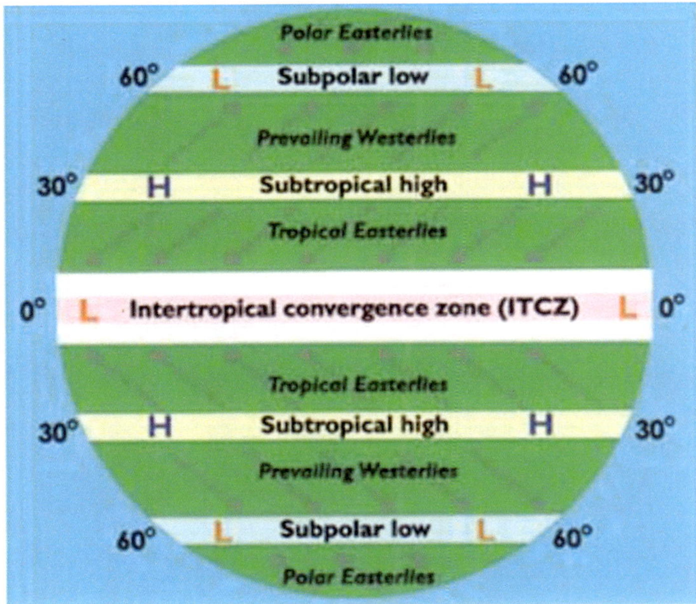

Fig. 5.8: Global Atmospheric Pressure & Wind Belts
https://drrajkumars.com/atmospheric-pressure/

ii) **Prevailing Westerlies:** These winds in middle latitudes between 30° to 60° blow from horse latitudes towards poles from west to east. These winds are predominately from southwest in the northern hemisphere and northwest in the southern hemisphere. The westerlies are generally strongest in the winter hemisphere and at times when the pressure is lower over the poles. They are weakest in the summer hemisphere when the pressures are high over the poles. The westerlies play a prominent role in carrying warm, equatorial waters and winds to the western coasts of continents, especially in the southern hemisphere because of vast area covering oceans.

iii) **Tropical Easterlies:** Near the equator, northeasterly winds from the northern hemisphere and south easterly winds from the southern hemisphere converge. This region is therefore known Inter Tropical Convergence Zone (ITCZ). The convergence of air facilities ascent of air upwards in the equatorial regions. These winds are popularly known as 'Trade winds' also.

5.23 General Circulation Pattern

The average circulation of air over the globe to describe the transport of energy consists of three cells names Hadley cell, Ferrel cell and Polar cell (Fig. 5.9).

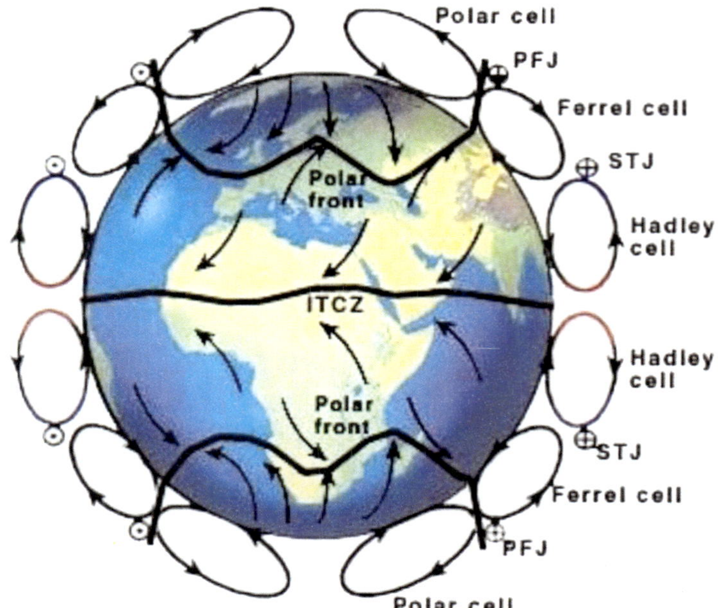

Fig. 5.9: General Circulation Pattern

https://www.metlink.org/resource/in-depth-the-global-atmospheric-circulation/

i) **Hadley Cell:** In the Hadley cell, air rises upwards into the atmosphere near the equator, flows towards the poles above the surface of the earth, descends down to the earth's surface at the sub-tropics and flows back towards the equator. The flow of air occurs because the sun heats air at the earth surface at the equator.

ii) **Ferrel Cell:** The Ferrel cell experiences a reversed flow pattern compared to Hadley cell between 30° to 60° latitudes. The air flows from sub-tropical latitude to sub-polar latitudes at the surface, rises up along the polar front and moves towards the sub-tropical regions.

iii) **Polar Cell:** The polar cell has similar circulation pattern as the Hadley cell. Though the air at higher latitudes is much cooler and drier than at the equator, the air masses at 60° latitude are still warm enough for convection to take place. The warm air rises up at 60° Latitude and moves towards the poles. When the air reaches the polar areas, it gets cooled considerably and sinks to finally move back towards 60° Latitude along the earth's surface.

5.24 Jet Streams

Jet streams are very strong air currents found in the atmosphere and are located near the tropopause. There are two main jet streams in the northern hemisphere. Although jet streams also do exist in southern hemisphere, they are strong in northern hemisphere. The wind speeds may vary between 100 to 200 miles per hour. There are two main jets known as Polar jet and Sub-tropical jet.

i) **Sub-Tropical Jet stream:** It is a belt of strong upper-level winds lying above the regions of sub-tropical high pressure. It travels in lower latitudes and at slightly higher elevations due to increased height of tropopause. In Tropics, an easterly jet is sometimes found at upper levels when a land mass extends pole wards.

ii) **Polar Jet Stream:** The polar jet stream is associated with the boundary between higher latitude cold and lower latitude warm air, popularly known as polar front. The average position of the polar jet stream changes seasonally. Its winter position tends to be at a lower attitude than during summer. Because of north-south temperature contrasts are greater in winter than summer, the polar jet stream winds are faster in winter than in summer.

5.25 Monsoon Winds

Monsoons are generally defined as a system of winds characterized by seasonal reversal of its direction. Monsoons are produced by the differential heating of the land and ocean.

For the same amount of heat received, the land surface gets heated more than the water as the specific heat of soil is much less than of water. The specific heat is defined as the amount of heat required to raise the temperature unit mass of a substance by 1°C. For the same amount of heat received, the land surface gets heated more than the water as the specific heat of soil is much less than of water. The specific heat is defined as the amount of heat required to raise the temperature unit mass of a substance by 1°C. The specific heat of soil is about 0.2 Cal/gm/°C and the specific heat of water is 1.0 Cal/gm/°C. Therefore, one calorie of heat supplied to one gram of soil increases its temperature by 5° whereas the temperature of 1 gm of water increases by only 1°C. Therefore, the land gets heated more than the adjoining seas/oceans. This results in high pressure over sea and low pressure over the land causing movement of air from ocean towards the continent. During winter season, high pressure prevails over continents and low pressure prevails over the oceans. Therefore, there will be movement of air from continent to ocean exactly in the opposite direction. Therefore, seasonal winds known as monsoons are formed with a reversal in the wind direction by 180° as the season changes. Thus, it can be clearly seen that weather and climate in any geographical region is governed by the distribution of atmospheric pressure and prevailing wind systems to a considerable extent.

5.26 Wind

Air in horizontal motion is called wind. The direction from which wind is blowing is the wind direction or wind ward direction and the direction towards which wind is blowing is called Leeward direction. The wind direction is observed using an instrument known as wind vane. The wind velocity is the distance through which the air moves horizontally in unit time. As the wind may not be steady all the time, its velocity changes instantaneously. The wind speeds are measured using an anemometer by observing the wind over a period of three minutes and averaged for unit time. These days there are anemometers which record instantaneous wind velocity at a given time.

The wind speeds can be expressed in kilometers/hour or meters per second.

5.27 Variation of Pressure and Wind with Height

The atmospheric pressure decreases with increase in height as the length of the air column decreases with increase in height.

The wind speed gradually increases with increase in height. The wind speeds are comparatively less at lower levels due to frictional forces offered by the ground. As the height increases, the effect of frictional forces decrease and therefore the wind velocity increases with increase in height.

5.27.1 Thermal Wind: The thermal wind refers to the vertical shear of the geostrophic Wind.

5.27.2 Turbulence: Turbulence is an irregular motion of the air resulting from eddies and vertical currents. It is contrast to laminar flow.

5.27.3 Vorticity: Vorticity is a measure of the rotation of a fluid and is defined as the Curl of the Velocity.

5.28 Wind Rose

The wind rose is a pictorial representation of wind speed and frequency of direction for a given month or season or a year at a given place. The wind rose is useful to find the predominant direction of the wind and magnitude of the wind velocity. For construction of wind rose 16 directions are considered as follows:

North (N)	South (S)
North North East (NNE)	South South West (SSW)
North East (NE)	South West (SW)
East North East wind (ENE)	West South West (WSW)
East (E)	West (W)
East South East (ESE)	West North West (WNW)
South East (SE)	North West (NW)
South South East (SSE)	North North West (NNW)

The data recorded on wind speed and direction have to be analyzed to find the frequencies of occurrence of wind direction and the wind speeds recorded in each direction averaged. These are plotted on a circular diagram indicating the average wind speed from each of the directions by means of rectangular bar gives an idea on wind speed and the direction in which the bar is drawn gives the wind direction (Fig. 5.10).

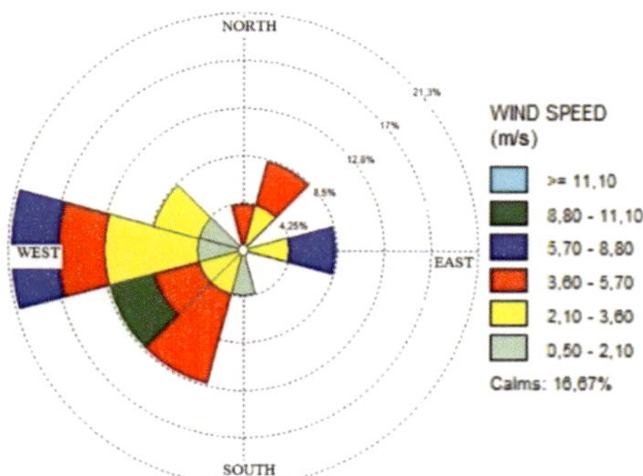

Fig. 5.10: Wind Rose Diagram
https://www.ppsthane.com/blog/wind-rose-beginners-guide

During some of the days the wind may be Calm (C). In such cases C is given within the circle as a numerical value. The wind direction may be highly variable during some of the days and the frequency of occurrence of variable wind (V) is also given within the circle. The wind roses are quite useful in designing shelterbelts and wind breaks to prevent soil erosion. In areas wherever irrigated crops are grown in isolated patches, wind breaks and shelterbelts can be used to prevent rapid depletion of soil moisture due to advection.

5.29 Atmospheric Waves

Atmospheric wave is a periodic disturbance in the fields of atmospheric variables such as surface pressure, temperature, wind velocity etc which may propagate (travelling) or not (stationary). There are two types of atmospheric waves that are important on earth and other planets.

5.29.1 Gravity Waves and Planetary Waves

The gravity waves are ripples in space and time generated by accelerated masses. The Planetary waves known as Rossby waves are a type of inertial wave occurring in rotating fluids. It is a large-scale perturbation of the atmospheric dynamical structure that extends coherently a full longitude circle. These waves naturally occur due to earth's rotation.

5.29.2 Land and Sea Breezes

The land and sea breezes are observed in areas near the sea coast. During day time, when the radiation is incident upon the earth's surface, the land gets heated more compared to the water in the sea (Fig. 5.11). The specific heat of the soil is approximately 0.2 Cal/gm/°C and that of water is 1.0 Cal/gm/°C. Therefore one Calorie of heat can raise the temperature of 1 gm of soil by 5°C and that of water by only 1°C. Therefore, the earth surface gets heated 5 times more than water in the sea. As a result, there will be high pressure over the sea and low pressure over the land. As wind blows from high pressure to low pressure, the cold air over the sea moves towards the land during the evenings usually after 3 pm or so. It is called sea breeze. During night times, the earth cools faster than sea water. For one calorie of heat lost due to long radiation, the temperature of 1 gm of soil decreases by 5°C whereas that of water decreases by only1°C. Therefore, around 4 am in the morning, there will be low pressure over the sea surface and high pressure over the land. There will be winds blowing from land towards the sea. It is called land breeze. Therefore; the land and sea breezes are formed due to differential heating between water in the sea and earth surface (Fig. 5.11.). At a height of about 1.0 km to 1.5 km above the earth surface, there will be high pressure over the land and low pressure over the sea during the evenings. As a result, there will be movement of air from land to sea when there is sea breeze. During early morning hours, the pressure at a height of about 1 to 1.5 km will be less over the land compared to sea. Therefore, the air moves from sea to land at that level.

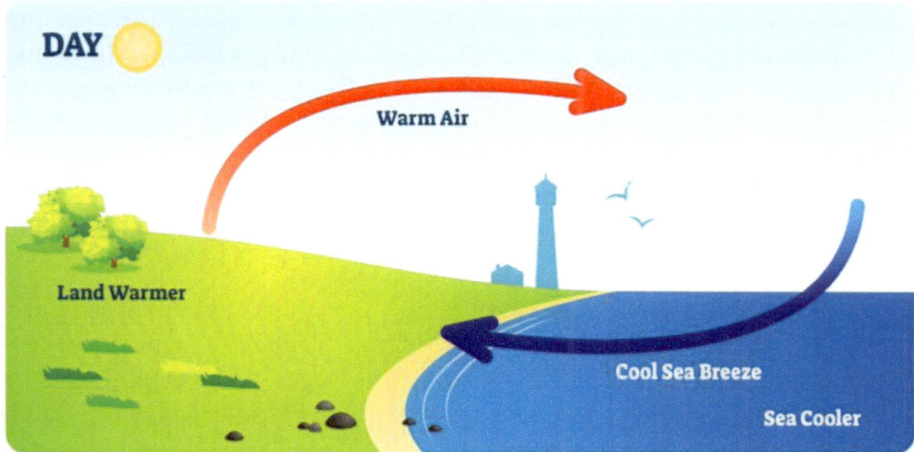

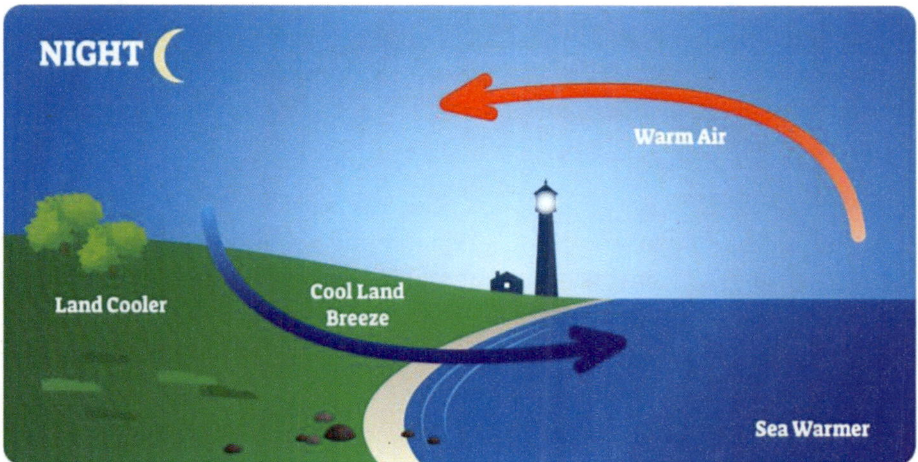

Fig. 5.11: Land and Sea Breeze Phenomena
https://www.science-sparks.com/what-is-a-sea-breeze/

5.29.3 Mountain Waves

The mountain waves are generated when strong wind blowing towards the mountain in perpendicular fashion are raised up over the mountains.

As the winds rise, they may encounter a strong inversion or stable air barrier over the mountains that cause the winds to be reduced towards the surface.

These mountain waves occur from a height about 1000 ft to 15000 ft. The mountain waves create lot of turbulence in the atmosphere and in aviation

5.29.4 Valley Winds

Mountains radiate heat at night and this makes mountains cool. At this time the valley is relatively warm. As a result, cold and heavy Winds blow down the mountain. The wind that blows at night towards the bottom of the valley is called mountain wind. This type of wind blows at night in Kullu and Kangra valleys in Himachal Pradesh. The mountain winds flow continuous from some time after Sun set till Sun rise. Mountain wind speed is highest in the night. Mountain winds are cool and heavy. Due to this wind, the bottom of the valley is covered with fog

Due to the slope of land, a special kind of flow is created in the mountain areas. During the day, the mountains are warmer than bottom of the valley and low pressure at the side of the mountain. For this reason, the wind that blows upwards through the mountain during the day is called valley wind. The valley winds blow during day time in the high lands. The air is warm and light. The valley winds are observed from Sunrise to Sunset. The wind speeds will be high in the afternoon. Due to the effect of valley wind, the upper part of the mountain region will be relatively warm which makes the weather pleasant and comfortable. Therefore, the population is relatively high in the upper part of such regions

5.29.5 T-Phigram

Entropy is a thermodynamic quantity representing the unavailability of systems thermal energy for conversion into mechanical work, often interpreted as a degree of disorder or randomness in the system.

A T-Phigram is one of the thermodynamic charts commonly used in weather analysis and forecasting. The name is evolved to describe the axes of temperature (T) and entropy. (Phi ϕ) used to create the plot. Usually temperature and dew point data from radio sondes are plotted at different heights (isobaric levels -hpa) on these diagrams. Calculations of convective stability are useful for the forecaster to predict unstable atmosphere to produce convective clouds likely causing thunder showers or stable atmosphere likely to produce stratiform clouds for steady precipitation and also possibility of fog occurrences.

The data required to plot T-Phi gram is collected using radio sonde equipment. It is equipped with sensors which respond to pressure, temperature and humidity and consists of two parts. One is known as sonde which carries sensors. The other part consists of radar on the ground. The radar has capacity to receive the radio waves transmitted by the sensors attached to the sonde. The sonde is attached to the balloon having hydrogen gas in it. When the balloon is released in the air, it starts emitting radio waves. The different sensors transmit radio waves of different wave lengths. The radio waves are received by the radar. The signals received at ground surface are converted into digital form. The data obtained on pressure, temperature and humidity are plotted on the tephigram. The drift in the balloon between different time intervals is useful in computing horizontal component of the wind.

5.30 Contour Charts used in Meteorology

The contour maps/charts are used by the meteorologists to find changes in weather variables (temperature, pressure etc) over larger areas and to estimate weather values at individual points where observations are not available. The following contour maps are most commonly used by the meteorologists

i) **Isobaric chart:** An isobar is line joining points having same atmospheric pressure on which isobars are plotted is called isobaric chart (Fig. 5.12). It is useful in identifying low pressure areas, high pressure areas, cyclones, depressions etc required for weather forecasting.

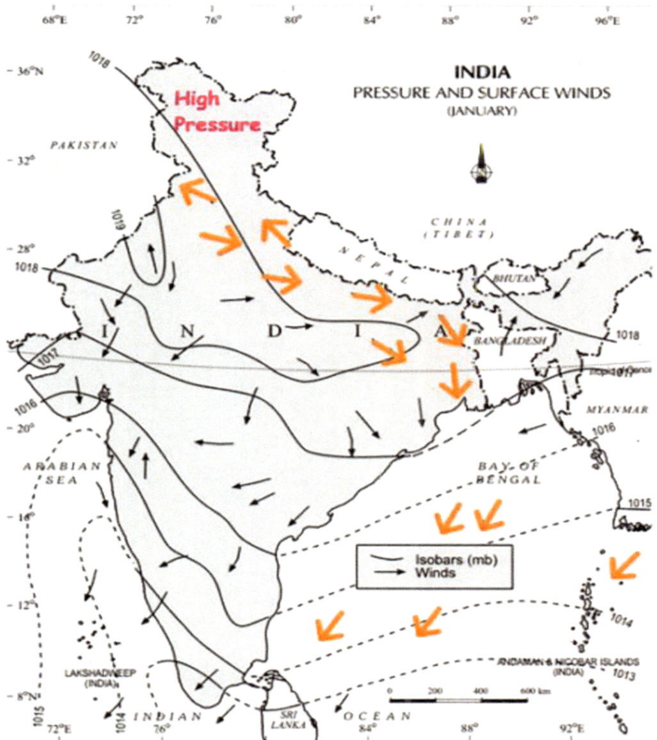

Fig. 5.12: Isobaric Map of India (January)
https://www.zigya.com/question/UjBWRlRqRXhNREEzT1RjMA==

ii) **Isothermal chart:** The line joining points having equal temperature is called isotherm. The map on which isotherms are plotted is called isothermal chart (Fig. 5.13). These charts are useful in identifying heat waves, cold waves, regions prone to changes in temperature etc.

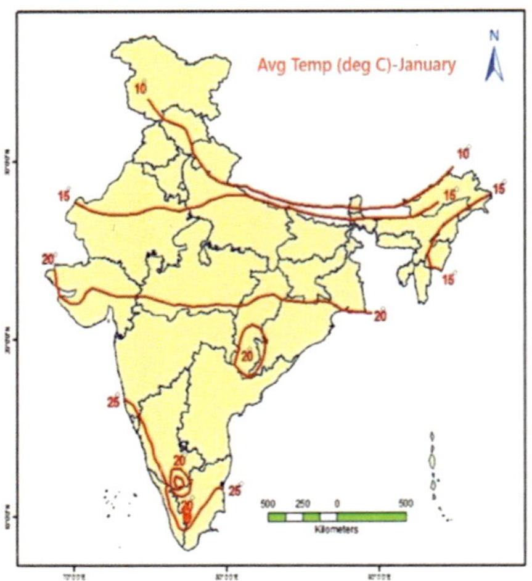

Fig. 5.13: Isothermal map of India (January)
https://www.researchgate.net/publication/321532699_A_Text_Book_on_Agricultural_Meteorology

iii) **Isohyetal chart:** The line joining points that received same amount of rainfall is called Isohyet. The map on which isohyets are plotted is called isohyetal chart. These charts are useful in identifying the distribution of rainfall in different regions (Fig. 5.14).

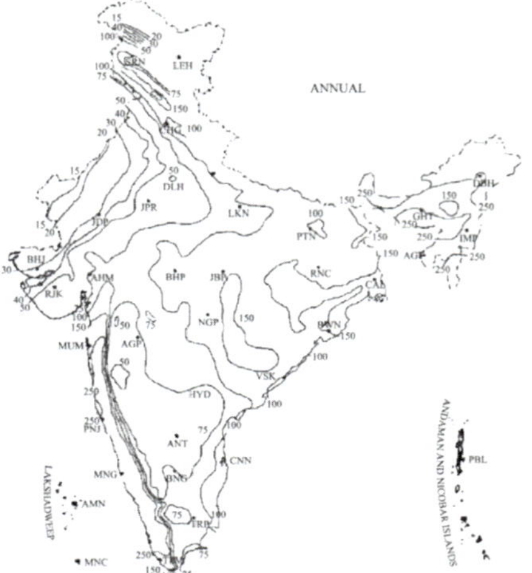

Fig. 5.14: Isohyetal map of India (Annual rainfall)
https://mausam.imd.gov.in/

iv) Isogon chart: Isogon is the line joining the points having same wind direction. The map on which isogons are plotted is called isogon chart. The chart used to find the changes in wind direction at regional level. (Fig. 5.15).

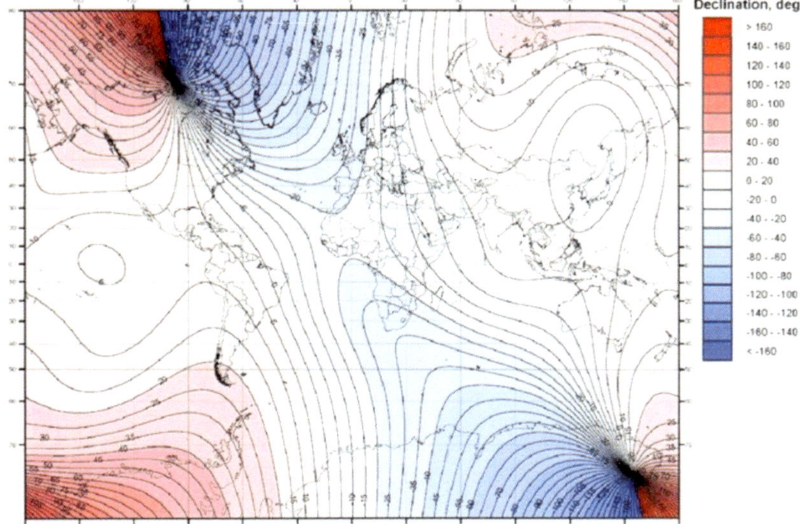

Fig. 5.15: Isogon map of World
https://www.researchgate.net/publication/253640760_Alexander_von_Humboldt%27s_charts_of_the_Earth%27s_magnetic_field_an_assessment_based_on_modern_models/figures?lo=1

v) Isotach: Isotach is the line joining the points having same wind speed. The map on which isotach are plotted is called isotach chart. An isotach chart is used to regions under the influence of strong winds (Fig. 5.16)

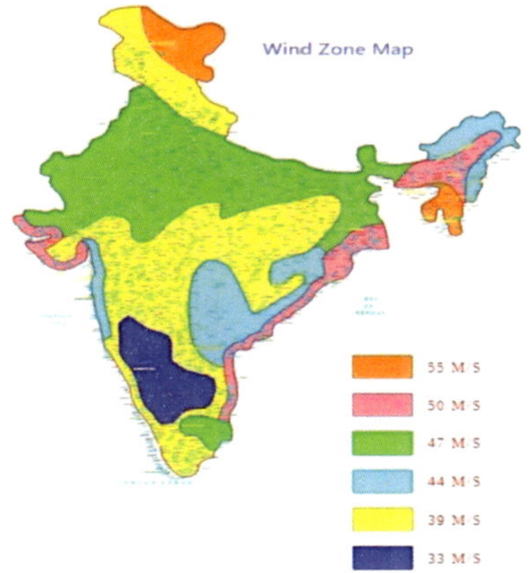

Fig. 5.16: Isotach (Wind Zone) map of India
https://dokumen.tips/documents/india-wind-zone.html

6

Moisture in the Atmosphere

Water exists in the atmosphere as water drops or ice particles suspended in the clouds and as well as water vapour. The water vapour content in the atmosphere is a very important observation as it indicates the extent to which air is either relatively dry or moist. The water vapour content in the air is measured as Humidity.

6.1 Absolute Humidity

Absolute humidity is expressed as weight of water vapour per unit volume of air. It is expressed in terms of grams of water vapour for one cubic meter of air. As the volume of air depends upon both pressure and temperature, it is not generally used. Suppose moist air rises vertically upwards from the earth's surface. As the parcel of air goes upwards, the atmospheric pressure decreases and therefore the air occupies greater volume. As a consequence, the value of absolute humidity decreases when the air moves upwards.

$$\text{Absolute humidity} = \frac{\text{Mass of water vapour in volume of air}}{\text{Volume of air}}$$

6.2 Specific Humidity

The specific humidity is defined as the weight of water vapour present in the unit weight of air. The units of measurement generally used for specific humidity are grams of water vapour present in Kilogram of air.

6.3 Mixing Ratio

The mixing ratio is defined as the weight of water vapour per unit weight of dry air. In general, the weight of moisture content per unit weight of dry air is comparatively less and the mixing ratio will be close to the value of specific humidity.

6.4 Vapour Pressure

The water vapour present in the air also exerts some pressure. The partial pressure exerted by the water vapour present in the air is called vapour pressure and it is generally expressed in millibars.

6.5 Saturated Vapour Pressure

The air will have certain capacity to hold water vapour at a given temperature. Therefore when the air is filled with water vapour to maximum of water holding capacity, the air is said to be saturated. Therefore, saturated vapour pressure is the pressure when the air is fully saturated with water vapour.

The extent of saturation of the air depends upon the air temperature and is directly related to it. As the temperature of air increases, it can hold more water vapour in it. When the temperature decreases, water molecules present in the air slow down and there is a possibility for the water drops to appear.

6.6 Dew Point Temperature

Air can hold more water vapour at high temperature than at lower temperature. When the air is not saturated, it will get saturated at some temperature when it is cooled and as a result, condensation of water vapour appears on surfaces. Therefore, the temperature at which the air gets saturated when it is cooled is called dew point temperature. The dew point temperature depends upon the moisture content of the air and it will be more when the air has more water vapour.

6.7 Relative Humidity

The relative humidity is defined as the percentage ratio of actual vapour pressure of the air to the saturated vapour pressure at a given temperature.

Therefore, relative humidity is a measure of the moisture in the air relative to the amount of moisture the air can hold. Without humidity, there will be no clouds or rain. Further, the water vapour absorbs heat and warms the temperature. Therefore, we feel warmer, when there is more moisture in the air. By knowing the air temperature and dew point temperature, the relative humidity can be calculated as

RH = (Actual vapour pressure/Saturated vapour pressure) × 100

The air temperature and dew point will be same when the air is fully saturated with water vapour.

Relative humidity indicates the amount of water vapour present in the air to the maximum amount of water vapour it can hold at a given temperature.

6.8 Saturation Vapour Pressure Deficit

The saturation vapour deficit is the difference between the saturated vapour pressure at a given temperature and the actual vapour pressure in the air. If the air is completely saturated, the saturation vapour pressure deficit is zero. If the air is more dry, the saturation vapour deficit will be more.

6.9 Condensation

When there is change in the phase of water vapour to liquid water, it is called condensation (Fig. 6.1). The amount of heat required to convert 1gm of water into water vapour at the boiling point of water is 540 cal/gm. If 1gm of water at room temperature needs to be converted into water vapour, it needs little more than 600 cal/gm. Therefore, when water vapour gets condensed into water droplets, it releases heat. The amount of heat required to convert 1gm of liquid to vapour at its boiling point is called Latent heat of vaporization. Similarly, the amount of heat required to convert one gram of solid into liquid at its melting point is called latent heat of fusion.

The latent heat of vaporization of water is 540 cal/gm and the latent heat of fusion of ice at its melting point is 80 cal/gm.

So when water is heated, it starts boiling at 100°C and further addition of heat to the boiling water will be used for converting as water vapour without further rise in the temperature of water beyond 100°C.

Similarly when heat is supplied to ice at its melting point of 0°C, it is used for melting of ice and for heating the water formed due to melting of ice. The transfer of heat involving change in the phase of substance like water is called latent heat. The amount of heat transferred to a substance like air without change of its phase is called sensible heat.

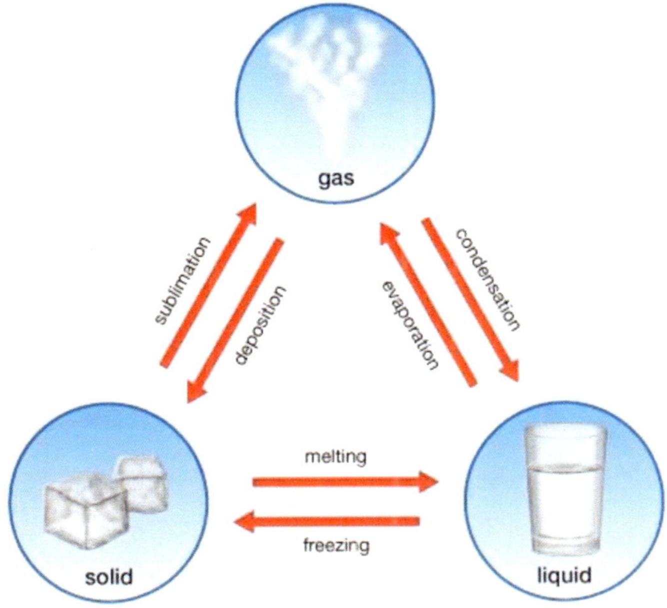

Fig. 6.1: Phase Changes of Matter
https://www.britannica.com/science/phase-state-of-matter

Therefore, the latent heat released due to condensation of water vapour in the atmosphere heats the air and gets converted as sensible heat. The sensible heat increases the temperature and makes it lighter than the adjacent air layers and creates a difference in the atmospheric pressure. The differences in atmospheric pressure will result in movement of air. In the process, there will be transformation of thermal energy into mechanical energy. The large-scale energy transformations in the atmosphere will lead to development of storms, cyclones etc.

Therefore, condensation is a very important process in the atmosphere and condensation takes place when the air is almost saturated. The air can be saturated either by adding water vapour to it at that temperature or by cooling it to dew point temperature. In the atmosphere, cooling is most common way for condensation to occur and formation of clouds. The air gets generally cooled when it rises vertically upwards or when it gets in contact with a cooler surface. Water vapour requires a nuclei other than that of gas for it to condense as liquid water drops. In the atmosphere, the dust particles, smoke particles and salt particles serve as condensation nuclei. The smoke particles are produced by burning of trees. The salt particles are produced by the action of waves in the ocean. Dust and soil particles are blown into the atmosphere by strong winds. These solid particles are also called aerosols and get washed out by precipitation which cleans the air.

6.10 Evaporation

Evaporation is the process by which water is converted from its liquid state to vapour state. The transfer of water from the earth's surface to atmosphere takes place due to evaporation of water from soils, water bodies and plants.

The rate of evaporation depends upon several factors as described below:

i) **Wind Speed:** The evaporation increases with increase in wind speed.

ii) **Temperature:** The evaporation increases with increase in temperature.

iii) **Humidity:** The evaporation increases with decrease in humidity. The evaporation increases with increase in saturation vapor pressure deficit.

The evaporation of water from the soil is termed as soil evaporation. The evaporation of water from the plants is called transpiration.

The combined loss of water due to evaporation from soil and transpiration from plants is called evapotranspiration. The evaporation is measured in terms of the depth of the water column that disappears due to evaporation and is expressed in millimeters.

6.11 Potential Evapotranspiration

The concept of potential evapotranspiration was developed by Thornthwaite (1948) and Penman (1948). Penman (1948) defined potential evapotranspiration as the maximum amount of water that can be lost due to evapotranspiration from a soil covered with short grass fully shading the ground when water is not limiting factor.

The concept of potential evapotranspiration is very useful in agriculture as it indicates the maximum demand for water when water is available in plenty.

6.12 Psychrometric Equation

A psychrometer is a hygrometer made up of two thermometers that are identical. One thermometer bulb is kept wet. When air passes through the bulbs of these thermometers, the dry bulb thermometer shows the temperature of air. When the air passes over the wet bulb thermometer, it evaporates the water over the wet bulb and cools it. Therefore, the wet bulb thermometer shows temperature less than that of dry bulb thermometer. Where the air is fully saturated and the relative humidity is 100% both the dry bulb and wet bulb thermometers record the same temperature. The equation used to calculate actual vapour pressure of the air based on dry bulb and wet bulb temperatures is called psychrometric equation as given below:

$$e = e_s T_w - 6.60 \times 10^{-4} (1 + 0.00115\, T_w)\, P(T - T_w)$$

where, e = actual vapour pressure of the air.

e_s = saturated vapour pressure at temperature T_d

T_d = dry bulb temperature

T_w = wet bulb temperature

P = Atmospheric pressure

When temperatures are below freezing point the value of 5.82×10^{-4} is used instead of 6.60×10^{-4}.

By knowing the actual vapour pressure at temperature T_d, the relative humidity can be calculated as

$$R.H. = \frac{e}{es} \times 100$$

6.13 Condensation Processes

Condensation is the process where water vapour becomes liquid water. Condensation happens one of two ways, either the air is cooled to dew point or it becomes so saturated with water vapour that it cannot hold any more water vapour. The condensation takes place around tiny particles (like salt or dust, smoke etc) in the atmosphere. These particles which facilitate condensation of water in the atmosphere are called condensation nuclei.

Therefore, the two processes by which condensation takes place are:

i) when the air is cooled to dew point

ii) when the air is fully saturated so that it cannot hold any further amount of water vapour

i) Dew: It is water in the form of small droplets that appear on thin exposed objects in the morning or in the evening due to condensation (Fig. 6.2). As the exposed surface gets cooled by radiating its heat, atmospheric moisture condenses at a rate greater than evaporation and water droplets gets deposited on the surface. The formation of dew takes place when the relative humidity is high and when the sky is clear.

Fig. 6.2: Dew Drops on Crop Plants
https://depositphotos.com/photos/dew-on-leaves.html

ii) Frost: Frost occurs when the temperature of the air in contact with ground is below the freezing point of water (Fig. 6.3). It is frozen dew. Frost occurs only when the temperatures are below freezing point.

Fig. 6.3: Frost on Crop Plants
https://www.cropsmart.com.au/should-i-spray-after-a-frost/

iii) **Haze:** Haze is light mist caused by particles of water or dust in the air which prevents visibility of distant objects as depicted in Fig. 6.4.

Fig. 6.4: Hazzy Conditions
https://www.hindustantimes.com/india-news/fires-dust-weather-plunge-capital-into-air-emergency-101667323926625.html

iv) **Mist:** Mist is a cloud of tiny water droplets suspended in the atmosphere at or near the earth's surface that restricts visibility above 1 km (Fig. 6.5).

Fig. 6.5: Misty Conditions
https://en.wikipedia.org/wiki/Mist

v) **Fog:** Fog is a cloud of small water droplets that is near the ground level and sufficiently dense to reduce horizontal visibility to less than 1000 meters (Fig. 6.6). In highly polluted air, the nuclei may be more to cause fog at humidity of even 95% or less.

Fig. 6.6: Foggy Conditions
https://www.ksla.com/2018/09/10/weather-or-not-how-does-fog-form/

vi) **Cloud:** A cloud is a mass of water drops or ice crystals suspended in the atmosphere, clouds form when the water condenses in the sky (Fig. 6.7).

Fig. 6.7: Clouds Over Crop Fields
https://en.wikipedia.org/wiki/Cumulus_cloud

vii) **Hail:** Hail is a solid form of precipitation consisting of solid ice that forms inside thunderstorm updrafts (Fig. 6.8).

Fig. 6.8: Hails Deposition
https://dictionary.cambridge.org/dictionary/english/hail

6.14 Bergeron-Findeisen Process

A theoretical explanation of the process by which precipitation particles may form within a mixed cloud (composed of both ice crystals and liquid water drops). The basis of this theory is the fact that the equilibrium vapour pressure of water vapour with respect to ice is less than with respect to liquid water at same sub-freezing temperature. Thus, within a mixture of these eyes and water particles the water vapour gets deposited on ice crystals rather than on liquid drops that would lose mass by evaporation. Upon attaining sufficient weight, the ice crystals would fall as snow and very lightly to become further modified by melting or evaporation before reaching ground. Operation of this process requires numerous small super cooled water drops which is a common feature in clouds between 0° and -20°C or below along with the small number of ice crystals. The crystals grow in size due to deposition of water vapour.

i) **Dry Adiabatic Lapse Rate:** The rate at which the dry air gets cooled with increase in height is called dry adiabatic lapse rate. The temperature of dry air decreases at the rate of 3°C/100 m height.

ii) **Moist Adiabatic Lapse Rate:** As a parcel of moist air moves upwards in the atmosphere, the water vapor in the moist air condenses and releases heat and warms air surrounding it. Therefore moist air cools slowly when it raises upwards at a rate of 6°C/1000 meters.

iii) **Atmospheric Stability:** The earth's atmosphere is a dynamic system and changes from time to time. Knowledge on atmospheric stability/instability is important for weather prediction. The stability of the atmosphere determines how conditions like pressure, temperature and wind speed vary with height in the atmosphere. The atmosphere is considered to be stable when changes in temperature, pressure and wind occur gradually with height. When air parcels are displaced slightly upwards or downwards, they tend to return to their initial position. When the atmosphere is stable, convection and vertical air movements are limited. It leads to calm and predictable weather conditions. There are three main types of atmospheric stability.

iv) **Absolutely Stable Atmosphere:** When the environmental lapse rate is less than dry and moist adiabatic lapse rates, the atmosphere is absolutely stable. When the lapse rate of the air parcel is less than that environmental lapse rate, the air parcel will be cooler than the environment even if it is pushed upwards. Therefore, it will not rise upwards and there will be no convection and sky will be clear.

v) **Absolutely Unstable Atmosphere:** When the environment lapse rate is more than the dry and moist adiabatic lapse rates, the atmosphere is considered as absolutely instable. When the air parcel is pushed upwards, it cools at a slower rate than the environment when it is unsaturated. Therefore, the air parcel will be warmer and lighter than the air in the environment and rises vertically upwards. It leads to formation of cumulus clouds due to strong vertical convections which may develop further as cumulonimbus clouds. Therefore, thunderstorms and rapid changes in weather can be expected.

vi) **Conditionally Unstable Atmosphere:** The atmosphere is said to be conditionally unstable if the environmental lapse is between dry adiabatic lapse rate and moist adiabatic lapse rate.

This means that the vertical rise of the air parcel depends upon whether the air is dry or saturated. If the air is saturated, the air parcel will be warmer and denser than the air in the environment and rises vertically upwards. It leads to formation of cumulus clouds which may further develop as cumulonimbus clouds resulting in the thunderstorms and rainfall.

vii) **Neutrally Stable Atmosphere:** Stable atmosphere will tend to resist vertical motion of air parcel. An unstable atmosphere will support vertical movement and lifting of air parcels. When the atmosphere neither resists nor supports vertical motion, it is said to be neutrally stable atmosphere. Vertical motion and instability are responsible for atmospheric turbulence and cloud formation.

6.15 Factors Controlling Atmospheric Stability

i) **Temperature:** The rate at which the temperature decreases with height of an air parcel determines stability. If the lapse rate is slow, the atmosphere is stable. A faster lapse rate makes the atmosphere unstable.

ii) **Moisture:** Higher moisture content in the lower atmosphere causes unstable conditions. Dry air makes the atmosphere stable.

iii) **Altitude:** The atmosphere is more stable at higher altitudes as the convection decreases at higher altitudes because of low temperatures.

iv) **Environmental Lapse Rate**: When the environmental lapse rate is more than the adiabatic lapse rate unstable conditions prevailed.

v) **Air Mass Properties:** Warm and moist air mass causes more unstable atmosphere.

6.16 Indicators of Atmospheric Stability

If the smoke particles from chimneys rise slowly and spreads horizontally at higher levels, the atmosphere is stable. If there's smoke particles rise rapidly and dispenses rapidly, it is unstable atmosphere.

i) **Cloud formation:** Instable atmosphere, the clouds will be stratified and can be seen at lower heights. In unstable atmosphere, there will be vertical growth of clouds.

ii) **Wind speed:** Strong winds cause unstable atmosphere.

iii) **Haze and visibility:** In stable atmosphere, haze and visibility tends to be poor at the absence of vertical lifting of air.

7

Cloud Classification

Clouds play very important role on weather and climate. The formation of clouds takes place when water condenses in the sky. Clouds are visible accumulation of tiny water droplets and ice crystals in the Earth's atmosphere. The moisture is always present in the atmosphere though it may not be visible. Clouds form when air becomes cooler until the water vapour present in it condenses to liquid form. The condensation over oceans occurs around solid particles like salt, dust, smoke etc. suspended in the atmosphere. The type of clouds depends upon temperature and wind. The clouds are classified based on physical appearance and the height at which clouds are formed (Fig. 7.1).

7.1 Classification Based on Physical Appearance

There are four major types of clouds namely cirrus, cumulus, stratus and nimbus.

i) **Cirrus Clouds:** Cirrus clouds are formed at altitudes of 8000 to 12000m. They are detached thin clouds. They appear white and feathery.

ii) **Cumulus Clouds:** Cumulus clouds are generally formed at a height of 4000 to 7000m. They look like cotton lumps. They exist in patches and can be seen dispersed. These clouds will have flat base.

iii) **Stratus Clouds:** Stratus clouds are spread horizontally. These are stratified or layered covering large portions of the sky. The stratus clouds are formed due to mixing of air masses of different temperatures or due to loss of heat. The stratus clouds cause overcast conditions and will obstruct sun light.

iv) **Nimbus Clouds:** Nimbus clouds are usually formed at lower altitudes. These clouds obstruct sun light. Nimbus clouds initially cause heavy rainfall and thunderstorms.

v) **Nimbostratus Clouds:** The nimbostratus cloud is a grey multilevel, amorphous, and nearly uniform in appearance. The cloud usually produces continuous rain, snow or sleet. Lightning and thunder are not associated with these clouds.

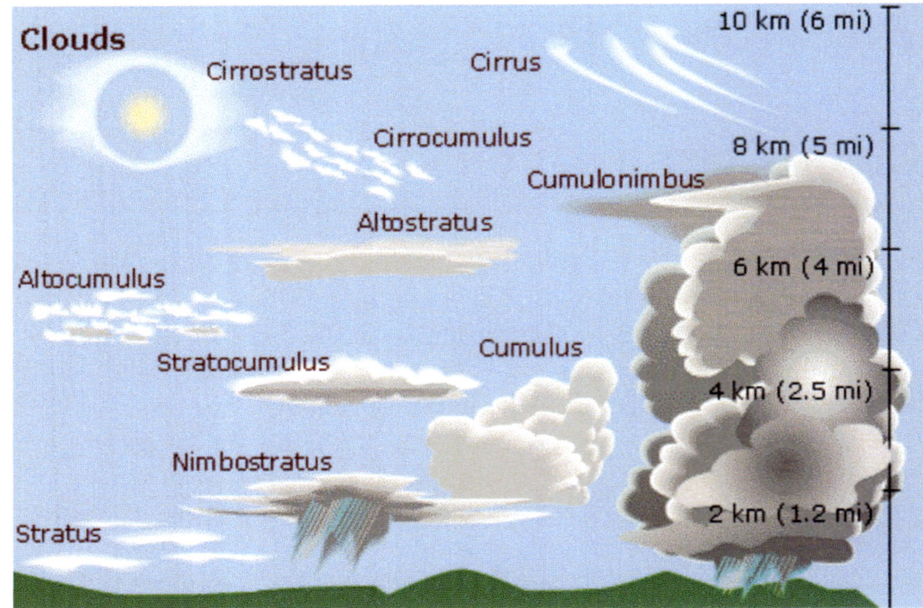

Fig. 7.1: Clouds Classification
https://salepeaket.live/product_details/8906348.html

vi) **Altostratus Clouds:** Altostratus clouds are large mid-level sheets. These clouds are composed of mixture of water droplets and ice crystals which are very thin allowing Sun to be seen weakly. These clouds are spread over very large area and are typically featureless.

vii) **Cumulonimbus Clouds:** Cumulonimbus clouds extend up to greater heights and look like towers with an anvil shape at the top. These are thunder clouds that produce hail, thunder and lightning. They look white in colour with light grey patches.

viii) **Altocumulus Clouds:** Altocumulus clouds are middle altitude clouds. These clouds are characterized by globular masses or rolls in layers or patches. The individual elements of the cloud look larger and darker compared to cirrocumulus clouds.

ix) **Cirrocumulus Clouds:** These clouds form at higher altitudes of 5 to 12 km. These clouds look white in color.

Classification of clouds based on Altitude

The clouds are classified depending on altitude as follows:

Classification	Cloud Types
High clouds	Cirrus, Cirrostratus, Cirrocumulus
Medium clouds	Altostratus, Altocumulus
Low clouds	Stratocumulus, Stratonimbus
Clouds with vertical development	Cumulus, Cumulonimbus

7.2 Thunderstorms

Thunderstorms are associated with lightning, thunder, strong winds, heavy rainfall and may produce hail storms (Fig. 7.2).

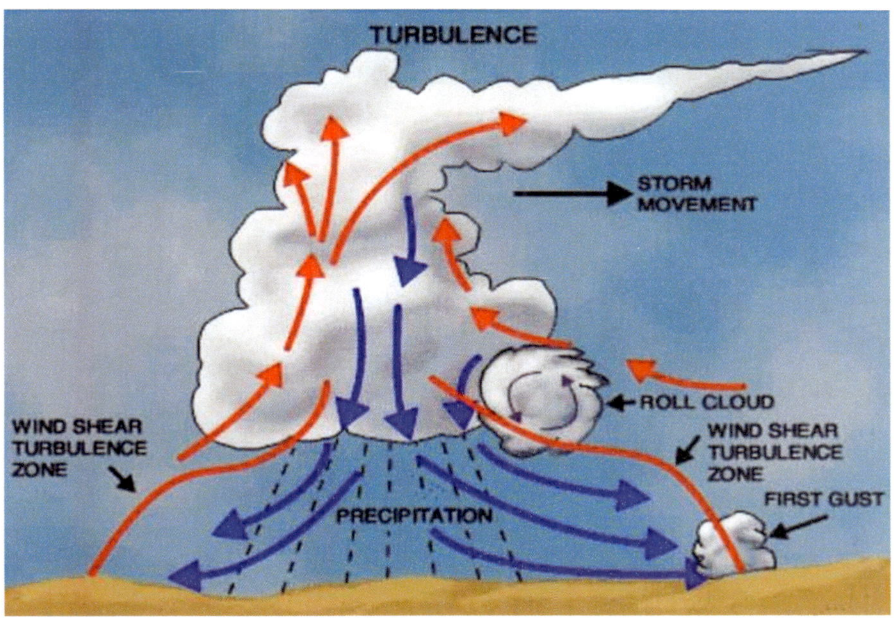

Fig. 7.2: Cross Section of a Thunderstorm
https://digitallylearn.com/thunderstorms-and-tornadoes-upsc-ias/

Some of the thunderstorms do not produce any rainfall but are associated with lightning and thunder. The thunderstorms occur due to cumulonimbus clouds which form when there is rapid rise or movement of warm and moist air. The cumulonimbus clouds may extend up to a height of 20km. The warm and moist air gets cooled and condenses when it moves upwards, resulting in formation of cumulonimbus clouds. When the rain drops fall through the clouds on way to earth surfaces, larger water drops form due to collision and reunion of small droplets with other droplets. When the raindrops are falling, it pulls cold air with it downwards. The down draft of air produces strong and gusty winds near the earth's surfaces.

The thunderstorms have following three stages

i) **Formative stage:** During daytime the earth gets heated by the sun radiation. It results in rapid upward movement of hot air as the hot air is lighter then cold air, the warm air condenses into a cumulus cloud in the presence of moist air. As long as the warm air below the clouds rises upwards the cumulus clouds will continue to grow.

ii) **Mature Stage:** As long as cumulus clouds grow, the water drops becomes larger and heavy. The rain drops fall after they become large and heavy when the rising air cannot hold them. When the rain drops start falling cool dry air starts entering the cloud. As the cool air is denser, it starts descending down as down drafts. The down draft forces the rain drops to move faster leading to more intense rainfall. The upward and downward drafts of the air lead to formations of cumulonimbus clouds.

iii) **Dissipating Stage:** The dissipating stage occurs when upward draft dominates downward draft in the clouds. Due to this domination, cloud droplets will cease to form. The cloud starts dissipating when the cloud disappears from top to bottom. Usually most of the thunderstorms last up to one or two hours. The hail produced during thunderstorms will cause severe damage to the horticultural crops and even kill livestock.

7.3 Dust Storms

Dust storms occur when strong, hot and dry winds blow dust and soil into the air. The strong winds move the dust and loose soil particles, often for many kilometres. The dust storms occur in the semi-arid and arid regions during the summer months after a period of drought. The dust storms may considerably reduce visibility to less than 1km.

During severe dust storms, more dust can get into the lungs. Dust irritates lungs and can trigger allergic reactions as well as asthma attacks. In desert regions, dust storms are commonly caused by thunderstorms or by strong pressure gradient which increases wind velocity. Dust storms cause soil loss from drylands and even remove soil fertility and reduce the production potential of the soil.

7.4. Precipitation

Water is most valuable resource for existence of life on earth. Precipitation is the water received from the atmosphere on earth's surface either in the form of solid or liquid water.

The precipitation occurs when water drops or ice particles grow big to overcome the upward force of lifting air. The precipitation occurs in different forms as mentioned below:

i) **Rain:** The water received from the atmosphere on earth's surface is called rain. The amount of water received through rain is called rainfall. The rainfall is measured in millimeters. The height of water column that gets accumulated due to rain on a horizontal surface of unit area (1 sq. cm) without being lost due to run-off is measured. Suppose the rainfall received is 50mm. It means that the water received due to rain on 1 square centimeter area is of height 50mm if it accumulates without being lost due to run-off.

ii) **Drizzle:** The rain received in the form of tiny water drops is called drizzle. The difference between rainfall and drizzle is intensity.

iii) **Hail:** The water received in the form of solid crystals is called hail. Hail is associated with rain as well as the ice melts partly before reaching earth's surface.

iv) **Snow:** Tiny crystals of ice that fall on earth are called snow. A crystal is solid substance with flat surface and sharp corners. Snow forms when tiny crystals joined together to become snowflakes.

v) **Sleet:** When falling snow reaches temperatures more than 0°C on its way to the ground, it melts. Sometimes, it will remain as liquid and reach the ground. In other cases, snow can partly melt before hitting the ground as frozen rain. It is called sleet.

7.5 Precipitation Processes

The precipitation mechanism can be explained by two processes:

i) **Bergeron Process:** The clouds contain ice crystals and super cooled water droplets. When an ice crystal collides with super cooled water, it freezes the water. This method uses two properties of water.

- **First property:** Cloud water droplets do not freeze at 0°C but remain liquid until -40°C as super cooled water. If the super cooled water is disturbed, it will freeze. It needs nuclei to freeze on. But freezing nuclei are rare in the atmosphere. Hence, further fall in temperature will turn some water droplets into ice and it will lead to all ice clouds with addition of an ice crystal to a super cooled water droplet.
- **Second property:** Over ice crystals the saturation vapour pressure is lower than over water. Between water and ice crystals, vapour pressure gradient is created. Ice crystals are formed at the expense of super cooled water. When these ice crystals get large enough, they

will begin to tumble out of cloud. These ice crystals melt and fall as rain before reaching the ground.

ii) **Collision-Coalescence Process:** This method is suited to clouds whose base does not exceed freezing point. These are warm clouds. These clouds have several cloud droplets of different sizes. Large drops expand at the expense of small water drops. So, the large water drops crash with the small water droplets and form part of it. The upward drafts and downward drafts repeatedly lift and lower cloud droplets in a huge cloud. The raindrops must have a diameter of 100 microns. The cloud droplets collide to generate a size of drizzle droplet. More collisions produce larger water drops and rain. Small droplets of uniform size may be present in clouds that do not produce rain. Because of small size of the droplets, a collision may not occur and these cloud droplets can fall slowly and uniformly without colliding. As a result, any clouds that lack required size of droplets may not produce rain.

7.6 Artificial Rain Making

Sometimes there may be clouds in the sky which may not produce rain either due to lack of condensation nuclei in the clouds or due to lack of sufficient cooling in the atmosphere. In order to produce rain artificially from such clouds, cloud seeding techniques are used. Cloud seeding is the process of spreading either dry ice(solid carbon dioxide) or more common silver iodide aerosols in the upper part of the clouds to stimulate the participation process and form rain.

Cloud seeding uses aircrafts to spray the condensation nuclei or silver iodide in the clouds.

There are different methods of cloud seeding as described below:

i) **Hygroscopic Cloud Seeding:** In this method salts are dispersed through flares or explosives in the lower portion of the clouds. The salt particles grow in size as water joins with them.

ii) **Static Cloud Seeding:** It involves spreading a chemical like silver iodide in the clouds. The silver iodide serves as condensation nuclei around which the moisture condenses and forms rain. The silver iodide makes rain bearing clouds more effective at dispensing their water.

iii) **Dynamic Cloud Seeding:** It aims to boost vertical air currents which encourage more water to pass through the clouds, translating into more rain. The process is considered more complex than static cloud seeding because it depends upon the sequence of events working properly.

7.7 Applications of Cloud Seeding

In agriculture, it might be used to save standing crops in drought stricken areas if there are suitable clouds present.

i) It can be used in catchment areas of river basins to augment water levels in the reservoirs.
ii) It can be used for suppression of hail and cyclone modification.
iii) It can also be used to tackle pollution of the atmosphere.

7.8 Air Masses

Air masses are large bodies of air with fairly uniform temperature and moisture characteristics. An air mass can be several 1000 kilometres across and extend upward to the top of the troposphere. Air mass is characterized by its surface temperature, environmental lapse rate and surface specific humidity. The air masses acquire the characteristics in the regions of their source. Insource regions, air moves very slowly and allows the air to acquire temperature and moisture from the region's land or ocean surface. Accordingly, types of airmasses are recognized as Maritime tropical (mT); Continental tropical (cT); Maritime polar (mP); Continental polar (cP) and Continental arctic (cA).

The air masses are classified based on the region where these are formed mainly into four types:

i) **Arctic:** The air masses are formed in the Arctic region and are very cold.
ii) **Tropical:** The air masses are formed in the low latitudes and are warm up to a moderate level.
iii) **Polar:** The air masses are formed in higher latitudes and are cold.
iv) **Equatorial:** The air masses are formed near the equator and are quite warm.

The air masses are classified based on whether they form overland or ocean.

i) **Maritime Air Mass:** Air mass formed over the water bodies are called maritime air masses.
ii) **Continental air mass:** The air masses forming overland areas are called continental air masses.

7.9 Fronts

The sharp boundary between two adjacent air masses is called front. The fronts are classified into three major types (Fig. 7.3).

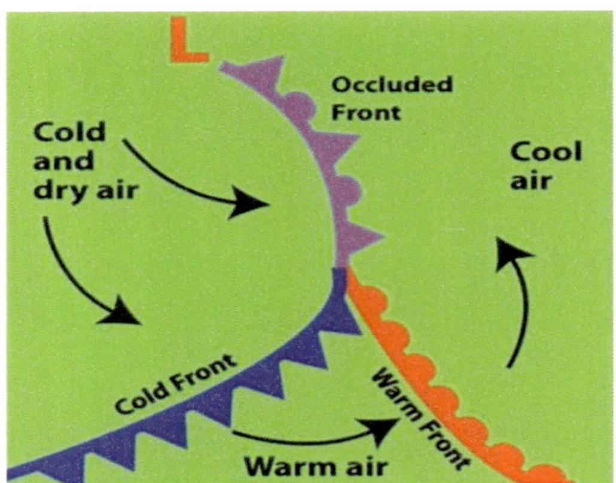

Fig. 7.3: Various Fronts

https://scied.ucar.edu/learning-zone/how-weather-works/weather-fronts

i) **Cold fronts:** When cold air invades warm air the boundary is a cold front. The temperature of the cold air mass is less and the air is more denser. The cold front remains in contact with the ground. As the cold front moves forward, it forces the warm air to rise above it. As the warm air rises upwards, it gets cooled adiabatically resulting in condensation of water and formation of clouds. If the warm moist air is unstable, the convection will be stronger.

ii) **Warm fronts:** In the case of warm front, the warm air moves into the region of cold air. The cold air mass remains in contact with the ground as it is cool and denser. The warm air mass is forced aloft as the air is lighter than the cold air mass. In this case the warm air moves over a long ramp with cold air below it. This motion creates stratus clouds which are large, dense and look like a blanket. These stratus clouds produce precipitation ahead of warm front. If the warm air is stable, there will be steady precipitation. If the warm air is unstable, there will be more convection leading to formation of cumulonimbus clouds which produce thundershowers.

iii) **Occluded front:** The occluded front forms when warm air gets caught between two cold air masses. The warm air mass rises as the cool air mass push and meet in the middle. The temperature drops as the warm air mass is occluded or cut-off from the ground and pushed upwards. After the occluded front passes, the sky is clear and the air is drier.

8

General Circulation

The general circulation of the atmosphere explains the formation of pressure belts and wind systems in the earth's atmosphere. It is known that the earth receives more solar radiation near the equator and gets heated more. Consequently, the air gets heated more and rises vertically upwards. As there is more convection of the air near the equator, the atmospheric pressure will be less. The warm air which moves upwards up to a height about 14 km near the equator gets cooled and move towards the poles.

The upper air gets gradually cooled when it reaches a latitude of 30°N in the northern hemisphere and 30°S in the southern hemisphere. At about 30°latitude the cold air descends to the earth's surface. Some of the cold air after descending to earth's surface moves towards the equator. The circulation of air between the equator and 30°N or 30°S latitudes is called Hadley cells (Fig. 8.1).

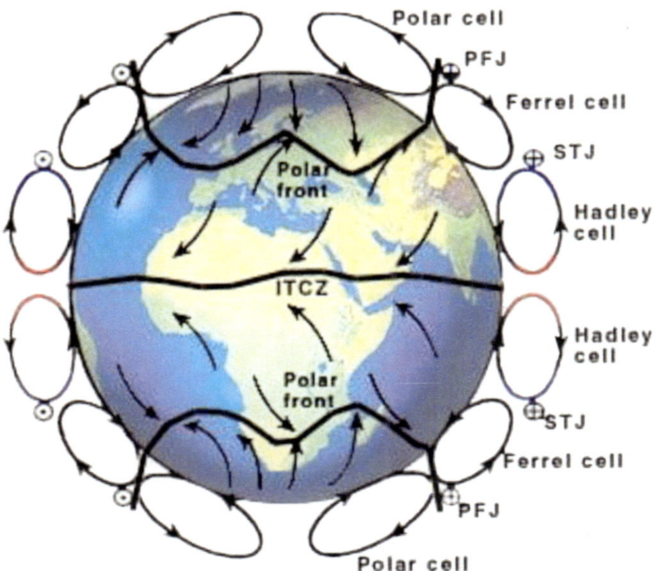

Fig. 8.1: General Circulation
https://www.metlink.org/resource/in-depth-the-global-atmospheric-circulation/

The winds blowing from 30° latitude towards the equator at surface level are called trade winds. These trade winds which move towards the equator in both northern hemisphere and southern hemisphere converge near the equator and the zone of the convergence of these winds is called Inter Tropical Zone of Convergence. As the convergence of air takes place, due to convection of heat as a result of high temperatures leads to vertical lifting of moist air. Therefore, the Equatorial regions receive more rainfall and there are tropical forests in this region.

The other part of cold descending air at 30° latitude moves towards the poles at surface level. After reaching about 60° latitude, the surface winds get warmer and are lifted upwards and gets cooled. The air then moves towards the equator and descends towards the earth at 30° latitude. Therefore there will be circulation of air from 30°N or 30°S at surface level to 60°N or60°S latitudes and towards the equator from 60° to 30° latitude at upper level. The circulations of air between 30° to 60° latitude are called Ferrel Cell.

The temperatures will be very less near the poles. Therefore, there will be high pressure near the poles. The air moves from poles towards the equator at surface level. After reaching 60° latitude, the air becomes warmer and gets lifted due to convection. The rising air at 60° latitude gets cooled after reaching certain height and moves towards the poles. As it moves towards the poles, it gets cooled and descends over the poles. The circulation of air between poles and 60° latitude is called Polar cell.

As the air coming from equator and Polar Regions descends at 30° latitude, there will be high pressure in this region and it is called Sub-tropical High Pressure. The clouds will be less in this region leading to less rainfall. Most of the deserts are formed in this region because of descending air and lack of convection in the atmosphere.

At 60° latitude, there is convergence of warm air at surface level due to the air coming from both polar and subtropical regions. There will be lifting of air due to convection. Therefore, this region experiences severe storms. The low pressure formed at 60° latitude is called Sub-polar Low Pressure system.

Due to rotation of earth and coriolis force, the winds blowing at surface level from poles to sub polar regions are deflected westwards and move eastwards. These winds are called polar easterlies.

The surface winds blowing from 30°to 60° latitude are called prevailing westerlies or these winds move towards east due to deflection by coriolis force.

The region around 30° latitudes in the northern hemisphere are called as horse latitudes due to absence of rain and light winds.

8.1 Walker Circulation

The Walker circulation refers to an east-west circulation of the atmosphere above the tropical pacific, with air rising above warmer ocean regions (normally in the West) and descending over the cooler ocean areas (normally in the east). Its strength fluctuates with that of southern oscillation.

8.2 Southern Oscillation

The atmospheric pressure difference between Tahiti in Indonesia and Darwin in Australia is known as southern oscillation.

8.3 Cyclones and Anticyclones

8.3.1 Cyclones

Cyclones are caused by atmospheric disturbances around a low pressure area associated with swift and destructive air circulation. The cyclones are characterised as low pressure areas with strong circulation of air in the anticlockwise direction in the northern hemisphere and clockwise direction in the southern hemisphere.

The cyclones are classified as extra tropical cyclones and tropical cyclones.

i) **Extra Tropical Cyclones:** The extratropical cyclones occur in temperature zones and high latitude regions. The extra tropical cyclones originate in polar regions.

ii) **Tropical Cyclones:** Cyclones that developed between Tropic of Cancer and Tropic of Capricorn are called tropical cyclones. These cyclones are known as typhoons in China Sea and Pacific oceans; hurricanes in West Indies islands, Caribbean sea and Atlantic region (Fig. 8.2).

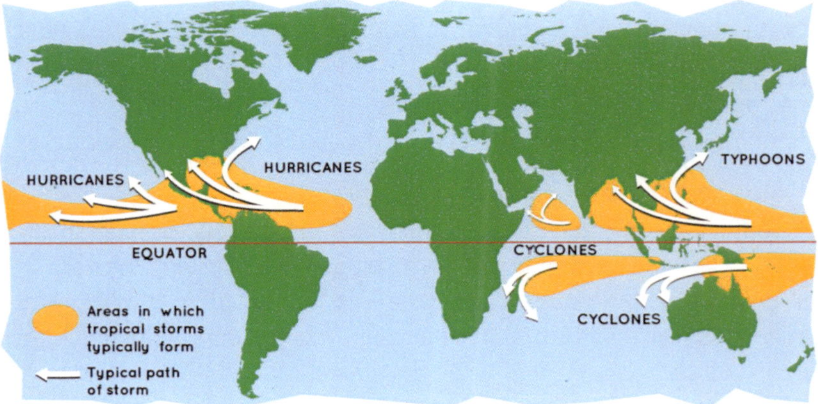

Fig. 8.2: Cyclones Distribution Over the Globe
https://commons.wikimedia.org/wiki/File:Hurricanes,_cyclones,_and_typhoons.jpg

Criteria of Indian Meteorological Department for classification of low pressure systems:

Type of Disturbance	Wind Speed(kmph)	Wind Speed(knots/hr)
Low Pressure	Less than 31	Less than 17
Depression	31 – 49	17 – 27
Deep depression	49 – 61	27 – 33
Cyclonic storm	61 – 88	33 – 47
Severe cyclonic storm	88 – 117	47 – 63
Super cyclone	More than 221	120

(1 knot = 1.85 kmph)

The cyclones are classified into five different types on the basis of wind speed and capacity to damage:

Cyclone Category	Wind Speed(kmph)	Damage Capacity
01	120 – 150	Minimal
02	150 – 180	Moderate
03	180 – 210	Extensive
04	210 – 250	Extreme
05	More than 250	Catastrophic

The development cycle of cyclones is divided into three different stages.

8.3.2 Formation and Initial Development Stage

The formation and initial development of cyclonic storm depends upon various conditions.

i) A warm sea (temperature more than 26°C to a depth of 60 metres) with abundant and turbulent transfer of water vapour to the overlying atmosphere by evaporation.

ii) Atmospheric instability encouraging formation of massive vertical cumulus clouds due to convection of rising air above sea surface.

8.3.3 Mature Tropical Cyclone

When tropical storm intensifies, the air rises in vigorous thunderstorms and tends to spread out horizontally at the tropopause level. Once the air spreads out, a positive perturbation of pressure is produced at higher levels. It accelerates the downward motion of air due to convection which is known as subsidence of air.

Due to subsidence air warms up by compression and a warm eye is generated.

The centre of the cyclone is called eye of the cyclone. The eye may have 3 different shapes:

i) Circular
ii) Concentric
iii) Elliptical

The main physical feature of a mature tropical cyclone in the Indian Ocean is a concentric pattern of highly turbulent giant Cumulus thunder cloud bands.

8.3.4 Modification and Decay Stage

When once the cyclonic storm crosses the Sea coast and enters the land area, it gradually gets weakened due to frictional forces offered by the ground.

In India, tropical cyclones occur in the months May-June and October-November. The disaster potential is particularly high during landfall in the north Indian Ocean (Bay of Bengal and Arabian Sea) due to accompanying destructive wind, storm surges and heavy rainfall occurs. Storm surges cause most damage as sea water intrudes in low lying area of coastal regions resulting in floods, damage to vegetation, soil salinity.

8.4 Anti-Cyclones

Anticyclones are high pressure centres. They are surrounded by closed isobars with pressure decreasing outwards. The winds blow in the clockwise direction in the northern hemisphere and anticlockwise direction in the southern hemisphere. The anti cyclones are associated with clear skies, sunshine and stable weather.

8.5 Tracking of Cyclones

The track of the cyclone is the line along which the centre of the cyclone moves. The cyclones rarely move along a straight line and takes a curved path depending upon the orientation of high pressure systems and prevailing wind directions. The path of each cyclone is continuously monitored to forecast the place where the cyclone may cross the Sea coast and to provide continuous warnings to the people living in coastal areas about the possible destructive activity associated with strong winds.

8.6 Western Disturbances

In north India, a series of low pressure systems travel from West to east during the winter season and these low pressure systems are called western disturbances. These western disturbances originate in Mediterranean region and bring some winter rains in north India (Fig. 8.3).

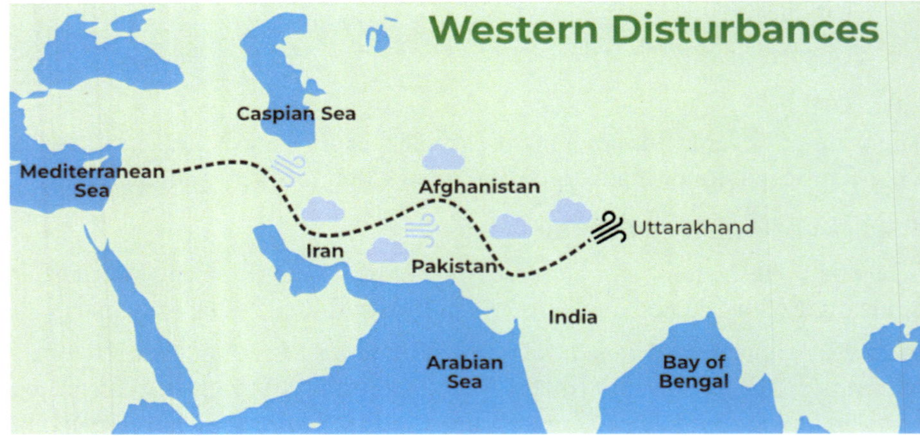

Fig. 8.3: Western Disturbance Route
https://www.geeksforgeeks.org/western-disturbances-in-india/

These are extra tropical storms. The low pressure systems carry moist air at upper levels and also associated with cold weather and even snowfall occurs in hilly regions of northern India. The formation of fog is also observed in association with western disturbances. These storms derive moisture from Mediterranean Sea and Black Sea. When the storms encounter the Himalayas in the Indian subcontinent, some of the moisture gets released as rain. The western disturbances usually occur during the months November to April. They are associated with merging of polar and subtropical air masses resulting in the formation of frontal boundary. The interactions between these air masses leads to cloud formation and precipitation. Before and after passage of western disturbance, temperature fluctuations occur. Warm air ahead of the system is replaced by cold air from the north, leading to fall in temperatures.

The precipitation received due to western disturbances is vital for crop growth during rabi season in Jammu and Kashmir, Punjab, Haryana and parts of Rajasthan. Additionally, the snowfall in the Himalayan regions ensure steady supply of water in rivers supporting irrigation and hydropower projects during dry season. On the average 4 to 5 western disturbances occur in a year. The indicators of western disturbances include cloudy skies, warm night temperatures and unexpected rain.

9

Climatic Classification

The division of earth's climates into worldwide system of contiguous regions, each one of which is defined by relative homogeneity of the climatic elements (Fig. 9.1).

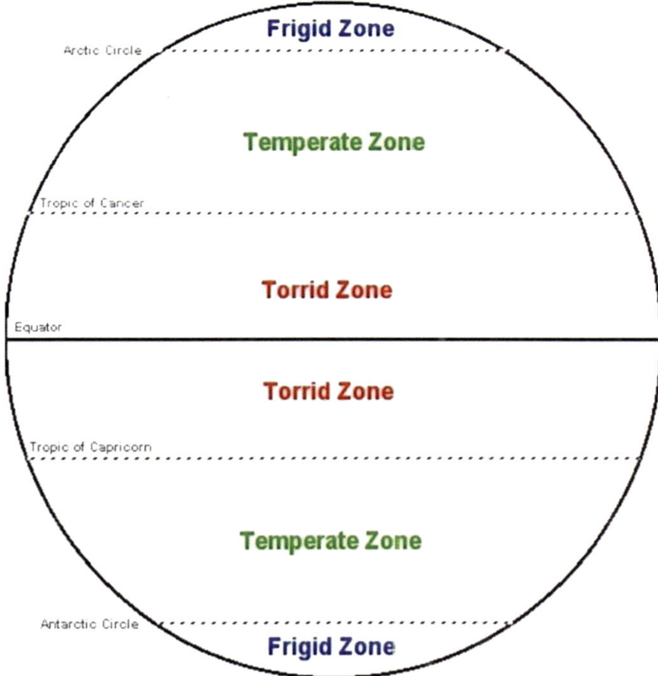

Fig. 9.1: Earth's Climates Classification – Greek (Aristotle)
https://www.insightsonindia.com/world-geography/physical-geography-of-the-world/climatology/world-climatic-regions/classification-of-climate-of-the-world/

9.1 Koppen's Classification

Koppen's climatic classification describes 5 major groups of climate A, B, C, D and E which are sub-divided into a total of 14 individual climate types, along with a special category of Highland (H) climate.

A - Tropical: Average temperature of coldest month is 18°C or more.

B - Dry climate: Potential evapotranspiration exceeds precipitation.

C - Warm temperature: The average temperature of the coldest month (of the middle latitudes) is higher than -3°C but below 18°C.

D - Cold climates: Average temperature for all months below 10°C.

E - Polar: low temperature and precipitation.

H - High land: Cold due to higher altitude of the region.

He introduced following sub-classifications in the 5 major groups of climates.

A_f: Tropical wet - No dry season.

A_m: Tropical monsoon - Monsoonal with short dry season.

A_w: Tropical wet and dry - winter dry season.

B_{sh}: Sub-tropical steppe - low latitude, semi-arid and dry.

B_{wh}: Sub-tropical desert - low latitude, arid and dry.

B_{sk}: Mid-latitude steppe - mid-latitude, semi-arid and dry.

B_{wk}: Mid-latitude desert - mid-latitude and dry.

C_{fa}: Humid sub-tropical - No dry season, warm summer.

C_a: Mediterranean - Dry hot summer.

C_{fb}: Marine west coast - No dry season, warm and cool summer.

D_f: Humid continental - no dry season, severe winter.

D_w: Sub-arctic - winter dry and very severe.

E_T: Tundra - No time summer.

E_F: Polar ice cap - perennial ice.

H: Highland - Highland with snow cover.

Thus, the climatic groups are sub-divided into types using small letters based on seasonality of precipitation and temperature characteristics.

The seasons of dryness are indicated by small letters f, m, w and s where:

f : no dry season.

m : monsoon climate.

w : winter dry season and,

s : summer dry season.

The abovementioned sub-groups are further divided depending up on the seasoned distribution of rainfall or degree of dryness or cold.

a : hot summer , average temperature of the warmest month more than 22°C.

c : cool summer, average temperature of the warmest month is less than 22°C.

f : no dry season.

W : no dry season in winter.

S : the dry season in winter.

G : Ganges type of temperature of annual march – hottest month comes before solstice.

h : average annual temperature below 18°C.

m : monsoon – short dry season.

The capital letters S and W are employed to designate the two sub-divisions of dry climate.

S : Semi-arid steppe.

W : Arid or desert.

The capital letters T and F are used to designate the two polar climates.

T : Tundra.

F : Ice cap.

Thus, the final description of different climates can be as follows:

Tropical Humid climates

A. Tropical wet	A_f	Wet all the year
Tropical savanna	A_w	Dry winter, wet summer
Tropical monsoon	A_m	Dry winter, very wet summer

Dry Climates

B. Sub-tropical desert	B_{Wh}	Hot desert
Mid-latitudedesert	B_{Wk}	Cold desert
Sub-tropical steppe	B_{Sh}	Hot semi-arid
Mid-latitudesteppe	B_{Sk}	Cold semi-arid

Mild Mid-latitude climates

Mediterranean	C_{sa}	Hot, dry summer
Humid sub-tropical	C_{fa}	Wet all year, hot summer
	C_{Wa}	Dry winter, hot summer
	C_{Wb}	Dry winter, warm summer

Severe Mid-latitude Climates

Humid Continental	D_{fa}	Cold winter, wet all year, hot summer
	D_{wa}	Dry winter, hot summer
	D_{wb}	Dry winter, warm summer

Polar Climates

Tundra	E_T	Polar tundra, no true summer
	E_F	Polar ice cap

High land Climates

High land climate	H	High altitude climates

9.2 Description of Climates

The major characteristics associated with different climatic groups and divisions given by Koppen are as follows:

i) **Tropical Humid Climates (A):** Tropical Humid climates are found about 15 to 25 degree latitude northwards and southwards of the equator. The temperatures in these zones remain above 18°C and the annual rainfall is usually above 1500 mm. Within this broad climatic zone, three minor climatic types exist.

ii) **Tropical Wet Climate (A_f)** will have precipitation around the year. The monthly variations in temperature in these regions are less than about 3°C. The extreme high humidity and surface temperatures in these regions cause cumulus and cumulonimbus clouds to form in the afternoons resulting in high amount of precipitation.

iii) **Tropical Monsoon Climate (A_m)** will get precipitation within 7 to 9 months of the year.

iv) **Tropical Wet and Dry Climate (A_w)** also called as Savanna climate experiences extended dry season during winter months. During wet season, the precipitation is generally less than 1000 mm and occurs mostly during summer season.

v) **Dry Climates (B)** has more climates evaporation and transpiration than total precipitation. The regions are found between 25 to 35° latitudes in both the hemispheres. There are four sub-divisions in this climatic zone.

vi) **Dry Arid Climates (B_w)** are habitats of xerophytic vegetation. The letters h and k are suffixed after B_w to indicate that these zones are located in sub-tropics and mid-latitudes respectively.

vii) **Dry Semi-Arid (B_s)** also known as steppe climate. These regions receive more precipitation compared to dry arid climate due to mid latitude cyclones.. The letters h and k are suffixed to indicate that these zones are located in sub-tropics and mid-latitudes respectively.

viii) **Moist Sub-tropical Mid-latitude Climates (C)** have warm and humid summers and mild winters. These climatic zones extend 30 to 50 degree latitudes in both the hemispheres. During summer months thunderstorms occur. There are three groups in this climate:

1. **Humid Sub-tropical Climate (C_{fa}):** Summers are hot and humid with frequent thunderstorms, winters are mild. Precipitation occurs during the winter due to mid latitude cyclones.
2. **Marine Climates (C_{fb}):** These climates are found on western coasts of continents. The climate here is humid with hot and dry summer. Winters are mild and heavy precipitation may occur due to mid latitude cyclones.
3. **Mediterranean Climate Zone (C_s):** The rainfall occurs mostly during winters. Summer rainfall is scanty. This climate zone includes Oregon and California states in US.

ix) **Moist Continental Mid-climates (D)** have warm summers and can be cool. Winters are very cold. The average temperature in the warmest month is more than 10°C while temperature in the coldest month is less than -3°C. These regions are prone to snow storms. These are three subdivisions in this group.

1. **D_w** :Dry winter
2. **D_s**: Dry summer
3. **D_f**: Precipitation round the year.

x) **Polar Climates (E) have low** temperatures round the year and warmest month having temperature less than 10°C. Greenland and Antarctica represent polar climates. There are two subdivisions in this group.

1. **E_T** : Polar Tundra in which soil is frozen.
2. **E_F**: Polar Ice caps – the surface is permanently covered with snow.

9.3 Drawbacks of Koppen Classification

i) It is based on mean monthly values of temperature and precipitation.

ii) Ignored precipitation intensity, daily temperature variations and rainy days.

iii) Ignored role of air masses.

iv) Not a genetic classification.

9.4 Thornthwaite's Classification

An American climatologist CW Thornthwaite proposed a climatic classification in 1931 and modified it in 1948 again.

i) **1931 Classification:** Thornthwaite introduced 2 indexes for his climatic classification.

1. **Precipitation effectiveness:** The ratio of precipitation (P) to the evaporation (E) Is termed as precipitation ratio.

$$\frac{P}{E} = \sum_{1}^{12} 1.5\left(\frac{\gamma}{T} - 10\right)^{10/9}$$

Where,

r = mean monthly rainfall in inches.

t = mean monthly temperature in °F.

On the basis of the $\frac{P}{E}$ ratio, he divided world climate into five types of moisture zones.

Humidity zone	**Vegetation**	$\frac{P}{E}$ **Index**
A (wet)	Rainforest	127
B (humid)	Forest	64 – 127
C (sub-humid)	Grasslands	32 – 63
D (semi-arid)	Steppe	16 – 31
E (arid)	Desert	less than 16

Thornthwaite further introduced four different subtypes in the above classification as follows:

r = adequate rainfall in all seasons

s = summer dry season

w = winter dry season

d = rainfall deficit throughout the year

Based on the above approach, he has identified 20 types of climate as follows:

A_r, A_s, A_w, A_d

B_r, B_s, B_w, B_d

C_r, C_s, C_w, C_d

D_r, D_s, D_w, D_d

E_r, E_s, E_w, E_d

2. **Thermal Efficiency:** The thermal efficiency is defined as positive departure from freezing point. The thermal efficiency $\frac{T}{E}$ ratio is calculated using the formula,

$$\frac{T}{E}\text{Index} = \sum_{i=1}^{12}\left(\frac{t-32}{4}\right)$$

where, t = mean monthly temperature in °F.The value of (t-32) indicates degrees of temperature above freezing point.

Based on above, he classified the world into 6 temperature regions.

Temperature Region	Climate	$\frac{P}{E}$ Index
A’	Tropical	127
B’	Meso-thermal	64 – 127
C’	Mico-thermal	32 – 63
D’	Taiga	16 – 31
E’	Tundra	1 – 15
F’	Frost	0

Applying the above criteria, he has identified 32 types of climate in the world.

AA'_r AB'_r AC'_r

BA'_r BA'_w BB'_r

BB'_w BB'_s BC'_r

BC'_r BC'_s

CA'_r CA'_w CA'_d

CB'_r CB'_w CB'_d

CC'_r CC'_s CC'_d

DA'_w DA'_d DB'_w DB'_r

DB'_s DB'_d

DC'_d

EA'_d EC'_d

D’ (Taiga)

E’ (Tundra)

F’ (Polar)

ii) 1948 Classification: In this classification, Thornthwaite used the concept of Potential Evapotranspiration. He defined Potential Evapotranspiration as the maximum amount of water that can be lost through evaporation and transpiration from the soil covered completely with vegetation when water is not a limiting factor. He develcped an equation to estimate potential evapotranspiration (PE_n) month-wise as

$$PE_n = 1.6\left(\frac{10T_n}{I_n}\right)^{\alpha}$$

Where, the mean monthly temp of the nth month (T_n) is between 0 to 26.5°C. I_n is the monthly heat index for nth month and α is constant.

In case the mean monthly temp (T_n) is more than 26.5°C, PE for the month is given by $PE_n = 415.85 + 32.34\ T_n\ - 0.43\ T_n^2$

The monthly heat index I_n for the nth month is given by the formula,

$$I_n = \left(\frac{t_n}{s}\right)^{1.514}$$

where, I_n = heat index for nth month

t_n = mean monthly temperature of nth month

The annual heat index I is a sum of monthly heat indices for the twelve months of the year.

$$I_n = \sum_{n=1}^{12}\left(\frac{t_n}{s}\right)^{1.514}$$

The constant α depends up on the annual heat index (I) and is given by,

$\alpha = 6.7510^{-7}\ I^3 - 7.7110\ I^2 + 1.791210\ I + 0.49239$

The value of PE calculated as given above is the potential evapotranspiration under standard conditions assuming that the duration of sunshine (Photoperiod) is 12 hours. So, he applied correction to find the potential evapotranspiration (PET) for each month using the formula,

$$PET_n = PE_n\left(\frac{d}{30}\times\frac{N}{12}\right)$$

where, PET_n = Potential evapotranspirations for nth month after correction for length of photoperiod.

PE_n = Potential evapotranspirations for nth month assuming the length of photoperiod is 12 hrs per day.

d = number of days in the month.

N = length of photoperiod during the month (Sunrise to Sunset).

The water surplus (S) during wet season and water deficit (D) during the remaining part of the year were estimated by comparing the precipitation values with monthly potential evapotranspiration.

The values S and D were used to calculate humidity index (I_h) and aridity index (I_a) as follows,

$$\text{Humidity index, } I_h = \frac{s}{PET} \times 100$$

$$\text{Aridity index, } I_a = \frac{D}{PFT} \times 100$$

The values I_h and I_a were used to further calculate moisture index MI using the formula,

$$MI = 100\left(\frac{s - 0.6D}{PET}\right)$$

The climates were classified based on the moisture index values as given below.

Climate Type	Moisture Index
A Perhumid	> 100
B_4 Humid	80 – 100
B_3 Humid	60 – 80
B_2 Humid	40 – 60
B_1 Humid	20 – 40
C_2 Moist Sub-humid	0 – 20
C_1 Dry Sub-humid	-20 to 0
D Semi-arid	-40 to -20
E Arid	-60 to -40

The Major climate types evolved as given above were further divided into subgroups based on values of humid index (I_h) and aridity index(I_a) as given below.

For Moist Climates (A, B_4, B_3,B_2,B_1, C_2)

	Sub-groups	Aridity Index
r	Little or no water deficiency	0 to 16.7
s	Moderate summer water deficiency	16.7 to 33.3
w	Moderate winter water deficiency	16.7 to 33.3
s_2	Large summer water deficiency	More than 33.3
w_2	Large winter water deficiency	More than 33.3

For Dry Climates (C, D, E)

	Sub-groups	Humidity Index
d	Little or no surplus water	0 to 10
s	Moderate summer watersurplus	10 to 20
w	Moderate summer water surplus	10 to 20

s_2	Large winter water surplus	> 20
w_2	Large winter water surplus	> 20

Thornthwaite also introduced the concepts of thermal efficiency based on the main annual values of potential evapotranspiration. The summer concentration of potential evapotranspiration was assessed by adding the potential evapotranspiration values for 3 consecutive months during which PET values were more. The total 3 months summer PET was calculated as percentage of the annual potential evapotranspiration.

$$\text{Thus, Summer Concentration} = \frac{\text{Sum of monthly PET for 3 hottest months}}{\text{Annual PET}} \times 100$$

Based on thermal efficiency and summer concentration the climates were identified as given below.

	Climate Type	Annual PET (cm)
A'	Mesothermal	114 and above
B'_4	Mesothermal	99.7 to 114.0
B'_3	Mesothermal	25.5 to 99.7
B'_2	Mesothermal	71.2 to 85.5
B'_1	Mesothermal	57.0 to 71.2
C'_2	Microthermal	42.7 to 57.0
C'_1	Microthermal	28.5 to 42.7
D'	Tundra	14.2 to 28.5
E'	Frost	< 14.2

Based on summer concentration of PET the following divisions were made.

Sub-division type	Summer concentration of PET in percent
a'	Below 48.0
b'_4	48.0 – 51.9
b'_3	51.9 – 56.3
b'_2	56.3 – 61.6
b'_1	61.6 – 68.0
c'_2	68. – 76.3
c'_1	76.3 – 88.0
d'	> 88.0

Thornthwaite and Mather develop the concept of water balance during the year 1955 using book-keeping procedure to estimate the surplus (S) and deficit (D) of water considering the soil moisture storage as 100 mm. They modified the formula to calculate moisture index (MI) as,

$$M.I. = \frac{100(s - D)}{PET}$$

Where, MI = moisture index

S = surplus of water

D = deficit of water

PET = Annual Potential Evapotranspiration

Based on the revised calculations of moisture index, the climates were classified as given below:

Climate Type	**Values of MI**
A Perhumid	Greater than 100
B_4 Humid	80 – 100
B_3 Humid	60 – 80
B_2 Humid	40 – 60
B_1 Humid	20 – 40
C_2 Moist Sub-humid	0 – 20
C_1 Dry Sub-humid	-33.0 to 0
D Semi-arid	-67.7 to -33.0
E Arid	-67.7 to -100

Both Koppen's and Thornthwaite classification were developed to classify world climates. However, there are several methods adopted for classification of regional climates depending upon the purpose for which climatic classification is required.

i) **Maritime Climate:** The maritime climate occurs in regions whose climatic characteristics are determined by their position close to the sea or ocean. In such regions are known for oceanic climates. About 72% of the earth is covered by water. The continents and oceans are unevenly distributed over the Earth's surface with northern hemisphere 65% of the continental area. The relative distribution of continental area and oceans play very important role in the climates that occur not only on sea, but also to varying degrees on land.

ii) **Continental Climates:** Continental climates experience significant annual variation in temperature (warm summers and cold winters). They tend to occur in middle latitudes 40 to 45°N with large land masses where prevailing winds blow over land bringing some precipitation and temperatures are not moderated by oceans. Continental climates occur mostly in northern hemisphere. A portion of the precipitation falls as snow and snow remains on the ground for at least one month or even more.

iii) Monsoonal Climates: Tropical monsoon climates have mean monthly temperatures above 18 in every month of the year and a dry season. Monsoons are seasonal winds with reversal in direction of flow by 180. The monsoons bring heavy rainfall during the months June to September. The monsoons always blow from cold to warm regions. The monsoon climate is found in the coastal regions of South West India, Sri Lanka, Bangladesh, Myanmar, Southwestern Africa, northeast and southeastern Brazil.

9.5 Hargreves Climatic Classification

Hargreves climatic classification is based on the concept of Moisture Availability Index (MAI) which is equal to the ratio of dependable precipitation to the potential evapotranspiration. He considered rainfall received during the month with 75 per cent probability as dependable rainfall.

$$\text{MAI} = \frac{\text{Rainfall that can be expected with 75 \% probability}}{\text{Mean monthly potential evapotranspiration}}$$

He classified semi arid climates as the regions with MAI more than 3 to 4 months in a year. If the MAI is more than 0.33 for only two months of a year, he classified those regions as arid regions.

9.6 Troll's Climatic Classification

The Troll's climatic classification is based on number of months with moisture availability. He defined the humid month as the month during which the mean monthly rainfall exceeds the mean monthly potential evapotranspiration. Based on the number of humid months in a year he classified the climates as follows:

Number of humid months	Climate type (s)
9.5 to 12	Tropical rainy climate
7.0 to 9.5	Tropical summer humid climates (dry summer)
7.0 to 9.5	Tropical winter humid climates (winter dry)
4.5 to 7.0	Wet dry tropical climates
2.0 to 4.5	Tropical dry climates (winter dry)
2.0 to 4.5	Tropical dry climates (summer dry)
Less than 2 months	Tropical descent climates

He considered regions with less than 2 humid months as arid climates and regions with 3 to 4 humid months as semi-arid climates.

10

Drought Climatology

Drought is a situation when the demand for water is much greater than the availability of water. Drought is a creeping phenomenon and it is experienced only after it has set in. The scarcity of water compared to demand will have greater impact on every activity of human life including failure of crops, food shortages, decline in ground water table, drying tanks, lakes or reservoirs, acute shortage of drinking water for humans and livestock shortage for cattle etc.

Drought will have very serious impact on economy of households as well as that of nation. There are five different types of droughts depending upon the nature of shortage of water

10.1 Meteorological Drought

Meteorological drought occurs when the water received through rainfall during a given period is much less than long period average rainfall of the area. The meteorological drought is associated with either failure of rain or due to long gap between the successive rainfall events. In India, meteorological drought occurs due to failure of south west monsoon or long breaks in south west monsoon rainfall. In Tamil Nadu the drought occurs due to failure of north east monsoon rainfall during months of October to December.

10.2 Hydrological Drought

The hydrological drought occurs due to decline in river water flows, depletion of water level in tanks, lakes, reservoirs, failure of rains in catchment areas of water bodies and fall in ground water levels (Fig. 10.1). Hydrological drought is also consequence of meteorological drought to a greater extent.

10.3 Agricultural Drought

Agricultural drought occurs when rainfall received and/or soil moisture storage are not adequate to meet the water requirement of crops and lead to either failure of crops or greater reduction in yields.

The agricultural droughts are classified into five types as described below:

1. **Early Season Droughts:** The early season drought occurs either due to delay in start of rainy season or early start of rainy season followed

by long dry spell. The early season drought may lead to withering of seedlings and inadequate plant stand, if there is early start of rain followed by long dry spell. If early season drought occurs due to late start of rainy season, the farmers will not be able to sow their regular crops and have to opt for alternate crops of short duration to compensate for the reduced length of rainy season.

2. **Mid-Season Droughts:** Mid-season droughts are associated with either failure of rains during vegetative phase of crop growth and or inadequate soil moisture availability to meet the water requirement of crop. In regions prone to mid-season droughts, soil moisture conservation strategies have to be followed by adopting suitable land configurations and use of mulches to prevent rapid drying of soil.
3. **Terminal Droughts:** The terminal droughts occur when there is early cessation of rainy season or failure of rains and inadequate soil moisture during the reproductive stage of crop growth. The terminal droughts will have greater impact on reduction of crop yields as the water requirements of crops is more during the reproductive stage of crop growth.
4. **Apparent Droughts:** Sometimes, it might so happen that some crops may be escaping drought while some other crops adversely affected. Such situation is called as apparent droughts. For example, in upland areas of sub-humid regions, crops like pigeon pea, maize, finger millet may not be affected by drought whereas as crops like paddy with higher water requirement may be affected by drought.
5. **Permanent Drought:** Permanent droughts occur in regions where even short duration crops might experience moisture stress even during the years with normal rainfall. Permanent droughts occur in arid and desert regions and regions experiencing permanent droughts are called drought prone areas.

   i) **Socio-Economic Drought:** The socio-economic drought occurs when the demand for an economic goods exceeds supply as a result of weather related to deficit in water supply. There are no uniform criteria for classification of droughts and will vary from one country to another. Therefore, the criteria used for classification of droughts was presented above

   ii) **Atmospheric drought:** Atmospheric drought is a condition when the air is dry, temperatures are high and there is no formation of rain bearing clouds in the region. Atmospheric drought increases the demand for water supplies.

10.4 Criteria Adopted for Declaration of Droughts

The government of India or even the state governments follow some standard procedures to identify and declare whether a particular region is affected by drought or not. The criteria adopted are based on scientific approach developed through research and recommended by the Ministry of Agriculture and Farmers Welfare, Government of India.

10.5 Broad Indices and Factors to Assessment of Drought During Kharif Season Rainfall Related Indices

i) **Rainfall Deviation:** The India Meteorological Department classified the rainfall deviation as given below.

Deviation from Normal Rainfall (%)	Category
+19 to -19	Normal
-20 to -59	Deficient
-60 to -99	Large Deficient
-100	No rain

ii) **Dry Spell**: Dry spell is a short period usually of 4 weeks (upto 3 weeks in light soils) of low rainfall or no rainfall. Thus, consecutive period of 3-4 weeks after the due date for the onset of monsoon with rainfall less than 50 per cent of normal in each of the weeks is defined as a dry spell. In sandy soil of arid areas (< 500 mm of rainfall) like Rajasthan, two consecutive weeks of dry spell may be considered along with rainfall deviation.

iii) **Standard Precipitation Index (SPI):** It expresses the actual rainfall as a standard departure with respect to rainfall probability distribution function. It can be positive, zero or negative. The positive values indicate wetness, and negative values indicate the dryness.

SPI value	Category
<-2	Extremely dry
-1.99 to -1.50	Severely dry
-1.49 to -1.00	Moderate dry
-0.99 to 0.00	Mildly dry

10.6 Impact Indicators

Remote sensing-based vegetation indices through regular monitoring of crop conditions/vigour over larger regions through satellite remote sensing technology and indicate the following;

i) **Normalized Difference Vegetation Index (NDVI)**: The NDVI values generally ranges from 0.2 to 0.6, the higher values of NDVI being associated with greater green leaf biomass.

ii) **Normalized Difference Wetness Index (NDWI):** Higher values of NDWI indicate more surface wetness. The ratio of NDVI/NDWI – deviation of -20 to -30 per cent represents moderate drought conditions and that of less than -30 per cent deviation from normal signifies severe drought.

iii) **Vegetation Condition Index (VCI):** It is expressed in percentage and higher values indicate good vegetation. Classification of vegetation condition based on VCI values are:

VCI value (%)	Vegetation condition
60 - 100	Normal
40 - 60	Moderate
0 - 40	Severe

10.7 Soil Moisture-Based Indices

Available soil moisture is a relevant indicator of drought especially in rainfed regions. The types of indices are commonly used as follows.

i) **Percent Available Soil Moisture (PASM):** The classification of drought based on PASM is given below:

PASM (%)	Agricultural drought class
76-100	No drought
51-75	Moderate drought
0-50	Severe drought

ii) **Moisture Availability Index:** Classification based on MAI (=AE/PE) i.e., the percentage ratio of actual evapotranspiration to the potential evapotranspiration is as follows:

MAI (%)	Agricultural Drought Class
76-100	No drought
51-75	Moderate drought
0-50	Severe drought

10.8 Rabi Season Drought

The crop production depends upon north east monsoon rainfall in major parts of Tamil Nadu and southern districts of Andhra Pradesh. For classification of droughts based on departure from normal rainfall, the rainfall data for the month of October to January is considered and the criteria used will remain the same as above.

11

Heat Waves and Cold Waves

Whenever the daily maximum temperature exceeds more than average values during summer season the heat wave will occur. Similarly, whenever the daily minimum temperature falls below the average values cold wave will occur. The India Meteorological department has developed the criteria to identify the occurrence the heat waves and cold waves.

11.1 Heat Waves

Heat wave is defined as the condition when the daily maximum temperature is 45°C or more. Based on departure from normal daily maximum temperature of 40°C onwards the heat waves are defined as follows:

Departure from normal temperature	Classification of heat wave
-1°C to +1°C	Normal
2°C	Above normal
3°C to 4°C	Appreciably above normal
5°C to 6°C	Moderate heat wave
7°C or more	Severe heat wave

Heat waves occur mostly in North and North western parts of India during summer more frequently.

During the period of heat waves, the sky will be very clear and air will be dry. High pressure prevails in the region. During the period of heat waves, people will be affected by severe sun stroke.

11.2 Cold Waves

The cold waves in plains are considered as and when the daily minimum temperature falls below 4-5°C in regions with mean daily temperature below 10°C. The following criteria is used as cold waves based on the deviation of actual minimum temperature from normal minimum temperature.

Departure from normal temperature	Classification of cold wave
-1°C to +1°C	Normal
-2°C	Below normal

-3°C to -4°C	Appreciably below normal
-5°C to -6°C	Moderate cold wave
-7°C or above	Severe cold wave

The cold waves occur in northern India after passage of western disturbances during the month of November to February. During the period of cold waves, thick fog may form late in the nights or in the early mornings. The cold waves may lead to occurrence of frost in the extreme parts of north India.

(Fig. 11.1). Schematic representations of the probability density function of daily temperatures. Dashed lines represent a previous distribution and solid lines a changed distribution. The probability of occurrence, or frequency, of extremes, is denoted by the shaded areas. In the case of temperature, changes in the frequencies of extremes are affected by changes (a) in the mean, (b) in the variance or shape, and (c) in both the mean and the variance

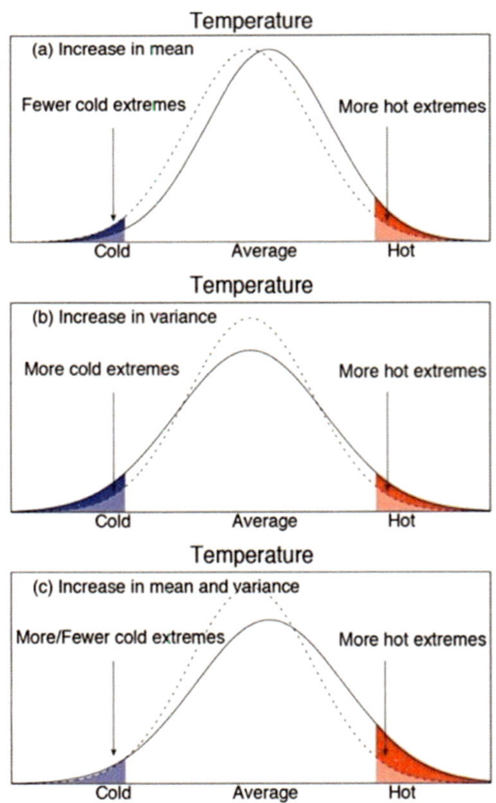

Fig. 11.1: Probability of Occurrence of Extremes Temp (Heat / Cold Wave)
https://www.ipcc.ch/report/ar6/wg1/chapter/chapter-11/

12

Global Weather Parameters and its Impact on Rainfall in India

The weather parameters that may occur outside India will also play a dominant role on weather conditions in India. Such weather parameters are called as global weather parameters and were found to play very important role on weather even in most parts of the world. During normal conditions in the Pacific Ocean, trade winds blow west along the equator, taking warm water from South America towards Asia. To replace that warm water, cold water rises from the depths-a process called upwelling. These global parameters are popularly known as El Niňo, La Niňa, Southern Oscillation, and ENSO (combination of El Niňo and Southern Oscillation).

12.1 El Niňo

El Niňo refers to the above average sea surface temperatures that periodically develops across east-central Pacific Ocean. During El Niňo, the surface / trade winds across the entire tropical Pacific Ocean are weaker than normal. Warm water is pushed back east, toward the west coast of the Americas (Fig. 12.1). El Niño means Little Boy in Spanish. South American fishermen first noticed periods of unusually warm water in the Pacific Ocean in the 1600s. The full name they used was El Niño de Navidad, because El Niňo typically peaks around December. For India, El Niňo is often linked with weak monsoon winds and dry weather, leading to reduced rainfall during the monsoon season. In India, generally drought years are El Niňo years but not all El Niňo years are drought years. El Niňo forms in the Pacific Ocean near South America around Peruvian coast.

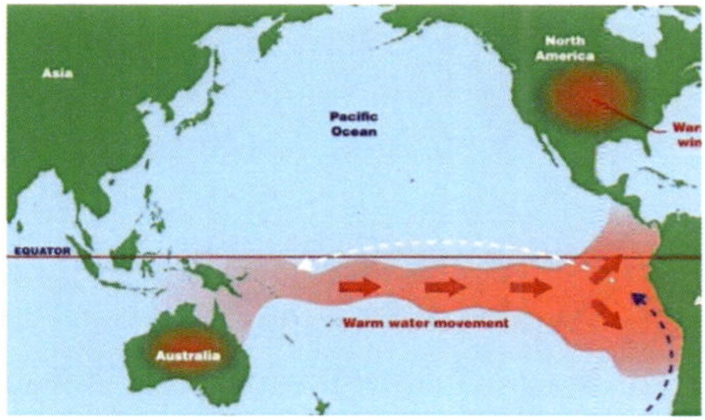

Fig. 12.1: El Niňo Conditions
https://scijinks.gov/

12.2 La Niña

La Niña means Little Girl in Spanish. La Niña is also sometimes called El Viejo, anti-El Niño, or simply 'a cold event.' La Niña has the opposite effect of El Niño. During La Niña events, trade winds are even stronger than usual, pushing more warm water toward Asia. Off the west coast of the Americas, upwelling increases, bringing cold, nutrient-rich water to the surface (Fig. 12.2). It refers to the periodic cooling of sea surface temperatures across the east central Pacific region. La Niña refers to the cooling of sea surface whereas El Niño refers to the warm phase of sea surface temperature. Generally, La Niña bring good rainfall during South West monsoon season.

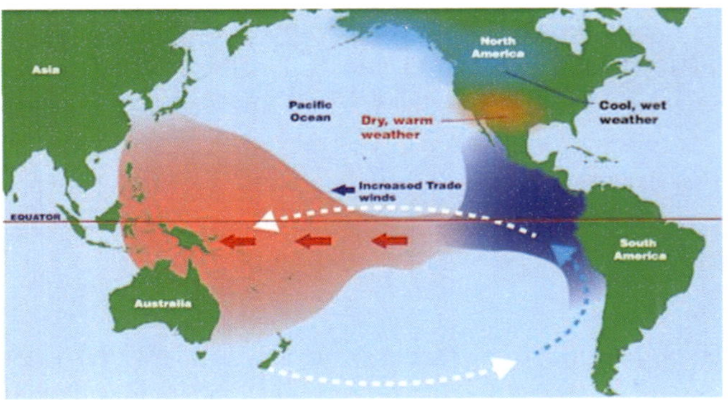

Fig. 12.2: La Niña Conditions
https://scijinks.gov/

i) **Southern Oscillation:** Southern Oscillation refers to the difference in atmospheric pressure between Tahiti in Vietnam and Darwin in Australia in equatorial Pacific Ocean. An important feature connected with Southern oscillation is the El Niño, a warm ocean current that flows past the Peruvian coast in South America. The changes in pressure conditions are connected with El Niño.

ii) **ENSO:** Since El Niño and Southern Oscillation are related, the two terms are often combined into single phase "El Niño- Southern Oscillation" or ENSO. The ENSO has impact on monsoon rain over north India and has become stronger during recent years, while weakening for the central region. The most prominent droughts in India, six of them since 1871 have been El Niño droughts. During recent years 2002 and 2009, the severe droughts were experienced in India during ENSO years.

iii) **Indian Ocean Dipole (IOD):** The Indian Ocean Dipole is defined on the basis of difference in sea surface temperature between a western pole in Arabian sea and eastern pole in eastern Indian ocean South of Indonesia (Fig. 12.3). IOD develops in the equatorial region of Indian Ocean from April to May peaking in October. When IOD is positive, winds over the Indian Ocean blow from east to West (from Bay of Bengal to Arabian Ocean). This results in Arabian sea (western Indian Ocean near African coast) being much warmer and eastern Indian Ocean around Indonesia becoming colder and dry. In the negative dipole year reverse happens making Indonesia much warmer and rainier. It was observed that positive IOD index often negated the effect of ENSO years like 1983, 1994, and 1997. The positive IOD (Arabian sea warmer than Bay of Bengal) results in more cyclones than usual in Arabian sea. The negative IOD leads to formation of cyclones in Bay of Bengal compared to Arabian sea.

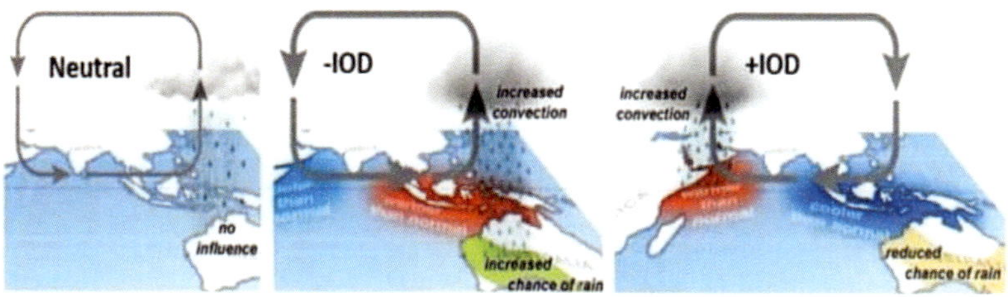

Fig. 12.3: Indian Ocean Dipole Conditions
https://www.slideshare.net/HarshitaPant6/indian-ocean-dipole-iod

vi) **Rossby Waves:** Rossby waves are formed when air from Polar Regions moves towards equator and the air from tropics move poleward. Due to the differences in temperatures between equator and Polar Regions, the transfer of heat from lower latitudes to higher latitudes takes place due to movement of air. Rossby waves are prominent in Ferrel cell. The concept of Rossby waves is used to explain the low-pressure cells and high-pressure cells that are important in changes of weather in middle and higher latitudes.

13

Climatic Features in India

India is one of the unique countries in the world with great diversity in climate. The annual rainfall varies from about 180 mm in North western part of India to almost 11,777 mm at Cherrapunji in Meghalaya. For a considerable amount of time, Cherrapunji had occupied the position of wettest place in India. However, off late, observations revealed that it is Mawsynram which receives the maximum amount of annual rainfall. Hence, Mawsynram with 11,872 mm annual rainfall is the wettest place in India. Both the places-Mawsynram and Cherrapunji-are located within 16 kilometers from each other in the East Khasi Hills district of Meghalaya. The maximum temperature at Churu in western Rajasthan is 50°C and the lowest minimum temperature at Leh in Ladakh region is as low as -35°C. In view of such great diversity in the distribution of rainfall and temperature in the country, it is necessary to have clear understanding on regional, seasonal and diurnal variation in different weather parameters (Fig. 13.1).

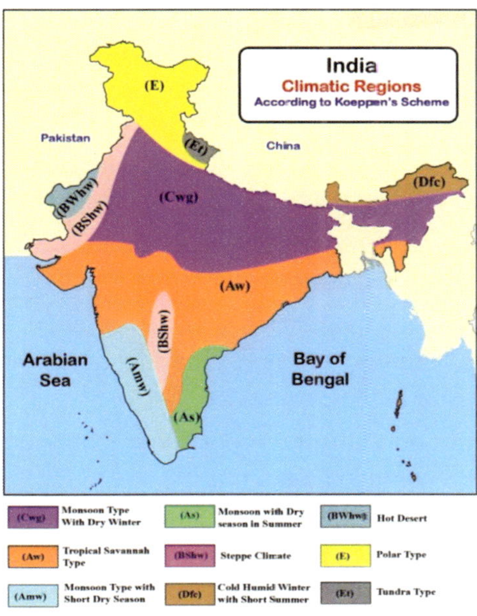

Fig. 13.1: Climate Regions in India (Koppen's Scheme)
https://geography4u.com/climate-zones-in-india/

Based on the distribution of rainfall and temperature, four major seasons are identified in India

Summer season: March to May

South west monsoon season: June to September

North east monsoon season: October to November

Winter season: December to February

The north east monsoon season is also referred as post-rainy season in most parts of India. North east monsoon rainfall is received in Tamil Nadu, Pondichary, southern districts of Andhra Pradesh, some parts of Karnataka and Kerala.

13.1 Temperature Distribution in India

The variation of temperature during the day (6 a.m. to 6 a.m. of the next day) is called diurnal variation. As the sun rises, the inclination of sun rays will be parallel to earths' surface and gradually the angle of incidence changes with time and will be perpendicular to the earths' surface at 12 noon. The earths' surface gets heated up and starts emitting long wave radiation. The constituents of atmosphere like water vapor, carbon dioxide, methane, nitrous oxide etc absorb the outgoing radiation whereas the abundant gases like oxygen and nitrogen cannot absorb it. The temperature of the air reaches maximum value around 2 p.m. and it is called as maximum temperature.

After 12 noon, the angle at which sun rays falling on earth decreases and will be parallel to earths' surface by sunset. As the earth emits long wave radiation, it gets cooled gradually and intensity of long wave radiation emitted by earth's surface decreases by about 4 a.m. in the early morning, the temperature decreases and reaches minimum value which is called as minimum temperature of the day.

i) **Mean daily temperature**: The average of maximum and minimum temperatures recorded during the day is called as mean daily temperature.

ii) **Diurnal range of temperature**: The difference between the daily maximum and minimum temperature recorded during the day is called the diurnal range of temperature.

iii) **Mean monthly maximum temperature**: The average of the maximum temperatures recorded in a month.

iv) **Mean monthly minimum temperature:** The average of the minimum temperatures recorded in a month.

v) **Mean annual maximum temperature**: The average of the mean monthly maximum temperatures from the month of January to December (Fig. 13.2).

vi) **Mean annual minimum temperature**: The average of the mean monthly minimum temperatures from the month of January to December.

vii) **Mean monthly temperature**: The average of the mean monthly maximum and minimum temperatures for the days in the month.

viii) **Mean annual temperature**: The average of the mean monthly temperatures for the month of January to December.

ix) **Annual range in temperature**: The difference between the highest maximum temperature recorded during the year and the lowest minimum temperature recorded during the year.

Fig. 13.2: Annual Temperature Distribution Over India
https://www.researchgate.net/publication/226241280_Sufferer_and_cause_Indian_livestock_and_climate_change/figures?lo=1

x) **Diurnal variation of temperature:** The variation in atmospheric temperature recorded during the day is called diurnal variation of temperature. In the areas closed to sea coast, the difference between the daily maximum and minimum temperature will be less. Therefore, the diurnal range of temperature is less. In typical land areas, far away from sea, the variation between the daily maximum and minimum temperatures will be more. Therefore, the diurnal range in temperature will be more. In coastal regions the relative humidity will be more in the air because of more moisture content of the air. Greater the water vapour present in the air, more the long wave radiation is absorbed. As water has higher heat capacity, its moist air cools slowly in the coastal regions and dry air cools rapidly in typical inland area away from the sea. Even the variation in temperature from month to month during the twelve months of the year will also be less in coastal regions compared to the inland areas away from the sea.

xi) **Temperature distribution during the month of January:** The winter season begins with the late November in North India and January month being the coldest month in most part of the country as sun shines vertically over the tropic of Capricorn in the southern hemisphere. During these periods the mean daily temperatures remain below 21°C in the northern plains and mountain regions in the north (Fig. 13.3).

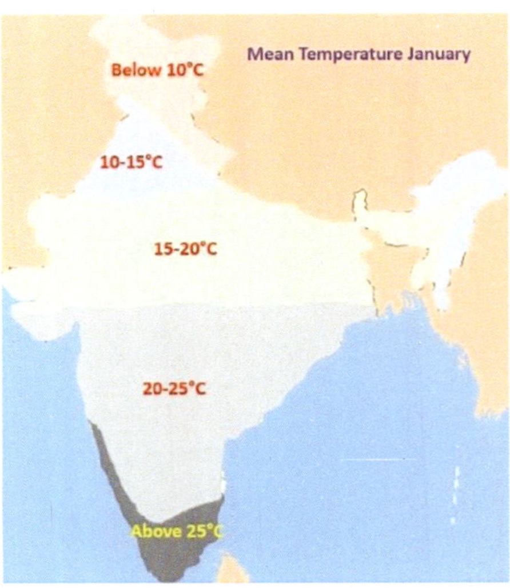

Fig. 13.3: Temperature Distribution in January Over India
https://prepp.in/news/e-492-seasons-in-india-geography-notes

The night temperatures fall below 0°C in the northern most parts of the country due to the passage of cold waves sometimes leading to occurrence of the frost. The temperature increases towards the southern part of the country. The mean daily temperatures will be around 25°C in southern most parts of the country. The mean daily temperatures will be 10°C or less in the northern most parts of the country.

xii) **Temperature distribution during the month of May:** Due to the apparent movement of sun towards the north, the temperatures start increasing. During the most part of the country, the temperatures reach higher values during the month of May and will reach up to 45°C in mid-May in the northern plains. The maximum temperatures will be around 35°C in the coastal regions along both east and west coast of India. The temperatures continue to be higher in the arid regions of northwest India even during the month of June. The maximum temperatures will be around 30 to 32°C in the north eastern parts of India.

13.2 Distribution of Atmospheric Pressure

During winter season, the atmospheric pressure will be high in northern plains of India due to low temperatures and low pressure prevails in the southern parts due to high temperatures (Fig. 13.4).

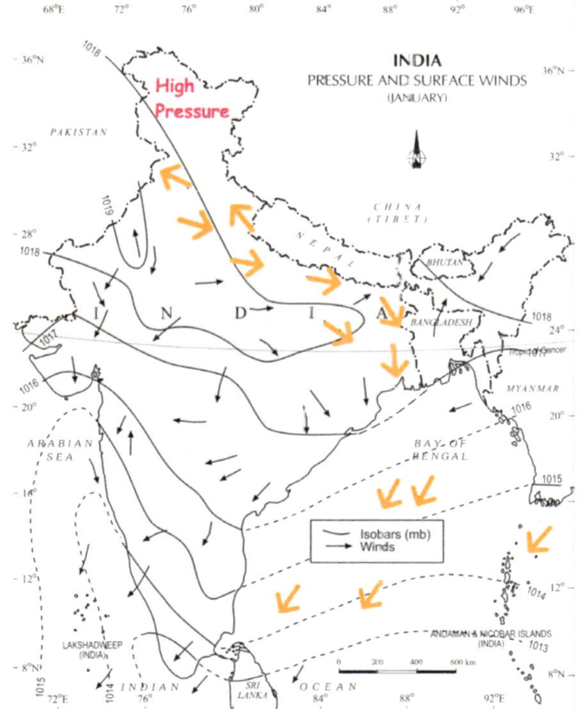

Fig. 13.4: Pressure During January Over India
https://www.civilsdaily.com/the-winter-season-january-february/

During summer season, low pressure prevails in the northern parts of India due to high temperatures and high pressure prevails over Indian ocean due to comparatively lower temperatures. The low pressure formed over north India during summer season is called thermal low.

13.3 Distribution of Annual Rainfall

The average annual rainfall in India is 1180 mm and it varies from about 180 mm at Jaisalmer in the north west India to almost 12,000 mm at Chirapunji in Kasi and Garo hills of Meghalaya (Fig.13.5).

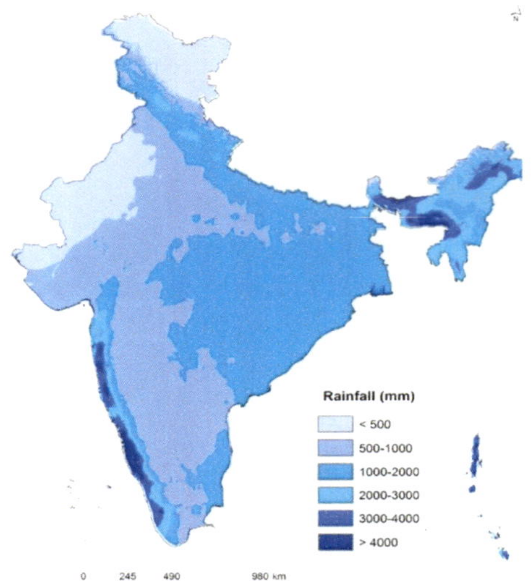

Fig. 13.5: Annual Mean Rainfall Distribution Over India
https://www.researchgate.net/publication/285020645_Nationwide_classification_of_forest_types_of_India_using_remote_sensing_and_GIS/figures?lo=1

i) **Regions with annual rainfall less than 500 mm**: The arid regions of western Rajasthan, Kutch and parts of Saurashtra region in Gujarat receive annual rainfall of less than 500 mm during the year.

ii) **Regions with annual rainfall between 500 to 1000 mm**: The rain shadow regions on the leeward side of Western ghats, western Uttar Pradesh, parts of eastern Rajasthan, western parts of Madhya Pradesh and most parts of Maharashtra, Central Karnataka, Telangana and northern parts Tamil Nadu, Punjab and Southern districts of Haryana and Delhi.

iii) **Regions with annual rainfall between 1000 to 1500 mm**: The average annual rainfall is generally between 1000 to 1500 mm in the regions along east coast, Orissa, coastal Andhra Pradesh, Chhattisgarh, eastern Madhya Pradesh, eastern Uttar Pradesh, Jharkhand, Bihar etc.

iv) **Regions with annual rainfall more than 1500 mm**: The average annual rainfall is more than on windward side of western ghats in Konkan region, Kerala, South Karnataka, west Bengal and north eastern India.

Most of these regions have extended rainy season from April to November in Kerala and from April to October in west Bengal and north eastern parts of India.

13.4 Main Rainy Seasons in India

Although the average annual rainfall in India is about 1180 mm, most of it is received during the southwest monsoon season. The long-term average annual rainfall for the period 1971-2020 in India during the Southwest monsoon season is about 870 mm (Fig. 13.6).

13.5 Distribution of Southwest Monsoon Rainfall

The period during the months June to September is called as Southwest monsoon season. The normal date of onset of SWM is 30thMay near Kerala coast. After it reaches Kerala coast, it gets divided into two branches popularly known as Arabian sea branch and Bay of Bengal branch. The Arabian Sea branch covers the western ghats, Mumbai, Gujarat and central India. These southwest monsoon winds of Arabian sea branch enter through Saurashtra to the northern plains. The Arabian Sea branch produces rainfall in peninsular India and central parts of India. Simultaneously the Bay of Bengal branch advances northwards and enters Bangladesh, Assam and adjoining states by 5th June.

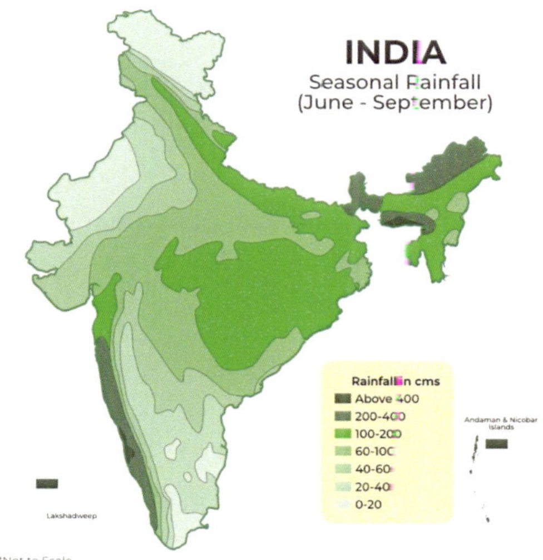

Fig. 13.6: SW Monsoon Rainfall Distribution Over India
https://www.geeksforgeeks.org/rainfall-in-india/

The topography and orientation of hills in Assam and Meghalaya obstructs the current in the north and east and results in deflection of monsoon winds towards the west under the influence of Himalayas. The Bay of Bengal branch enters the Gangetic plains as South easterly or easterly coast. Both Bay of Bengal and Arabian sea branch meet over north India along 80°E longitude. By 10^{th}June, the northern limit of monsoon passes Akola, Jagdalpur, Cuttack and Sabour. By 15th June it advances up to Varanasi and Gorakhpur. Further advance towards west is slow and reaches western Rajasthan by the end of first week of July.

The withdrawal of SWM commences from September 1. It withdraws from north west India from first September and completely withdraws from October 15, when north east monsoon sets in Tamil Nadu. The duration of south west monsoon varies from 2 months in North West India to 4.5 months Kerala and as well as in north eastern parts of India. However, the onset of south west monsoon may be early during some of the years and may be delayed as well near Kerala coast. The normal date of onset of southwest monsoon is 30th May near Kerala coast.

13.6 Onset of Southwest Monsoon and Advancement

Southwest monsoon normally sets in over Kerala on 1st June with a standard deviation of about 7 days. India Meteorological Department (IMD) has been issuing operational forecasts about two weeks in advance for the date of monsoon onset over Kerala from 2005 onwards. An indigenously developed state of the art statistical model with a model error of ± 4 days is used for the purpose. The 6 predictors used in the model are:

i) Minimum Temperatures over North-west India
ii) Pre-monsoon rainfall peak over south Peninsula
iii) Outgoing Long wave Radiation (OLR) over south China Sea
iv) Lower tropospheric zonal wind over equatorial southeast Indian Ocean
v) Outgoing Long wave Radiation (OLR) over Southwest Pacific Ocean
vi) Upper tropospheric zonal wind over equatorial northeast Indian Ocean.

For the year (2024), the southwest monsoon is likely to set in over Kerala on 31st May with a model error of ± 4 days.

The guidelines to be followed for declaring the onset of monsoon over Kerala and its further advance over the country are enlisted below:

i) Rainfall: If after 10th May, 60% of the available 14 stations enlisted*, *viz.* Minicoy, Amini, Thiruvananthapuram, Punalur, Kollam, Allapuzha, Kottayam, Kochi, Thrissur, Kozhikode, Thalassery, Kannur, Kudulu and Mangalore report rainfall of 2.5 mm or more for two consecutive days, the onset over Kerala be declared on the 2nd day, provided the following criteria are also in concurrence.

ii) Wind field: Depth of westerlies should be maintained upto 600 hPa, in the box equator to Lat. 10°N and Long. 55°E to 80°E. The zonal wind speed over the area bounded by Lat. 5-10°N, Long. 70-80°E should be of the order of 15 – 20 Kts. at 925 hPa. The source of data can be RSMC wind analysis/satellite derived winds.

iii) Outgoing Longwave Radiation (OLR): INSAT derived OLR value should be below 200 wm^{-2} in the box confined by Lat. 5-10°N and Long. 70-75°E.

Intense low pressure formed over Tibetan plateau due to more heating during summer season and high pressure in the South of Indian Ocean is the major cause of southwest monsoon. The normal date of onset of southwest monsoon at Kerala coast is June 1st. The monsoon winds beyond south Kerala progress in the form of two branches i.e; Arabian Sea branch and Bay of Bengal Branch. The progress of southwest monsoon under normal conditions over India can be seen in below Fig. 13.7.

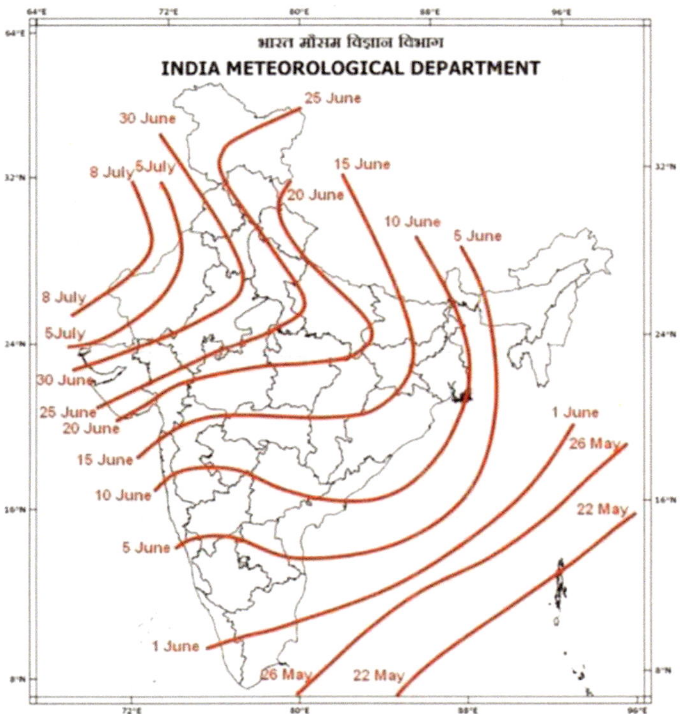

Fig. 13.7: Normal Dates of Onset of SW Monsoon Over India
https://internal.imd.gov.in/press_release20200515_pr_804.pdf

i) The Arabian Sea branch gradually advances northwards after onset of SWM on 30th May at Kerala coast. It reaches Mumbai by 10th June.

ii) The Bay of Bengal branch spreads rapidly over most of Assam. It reaches Kolkata by 7th June.

iii) After reaching foot hills of Himalayas, the Bay of Bengal branch is deflected westwards and moves towards Gangetic plain.

iv) Both the Arabian Sea branch and Bay of Bengal branch merge near New Delhi and form a single current as both branches reach Delhi about the same time.

v) The combined air current extends to Uttar Pradesh, Haryana, Punjab, Rajasthan and Himachal Pradesh. It enters Kashmir by mid-July and by that time it gets dried up.

vi) The Arabian Sea branch is more powerful compared to Bay of Bengal Branch.

vii) As the southwest monsoon winds start from Sea, they carry lot of moisture.

viii) The moist winds produce rainfall due to convection over land.

13.7 Retreating Southwest Monsoon

The following major synoptic features are considered for the first withdrawal from the western parts of NW India after 1st September:

i) Cessation of rainfall activity over the area for continuous 5 days.

ii) Establishment of anticyclone in the lower troposphere (850 hPa and below)

iii) Considerable reduction in moisture content as inferred from satellite water vapour imageries and tephigrams.

iv) Further withdrawal from the country is declared, keeping the spatial continuity, reduction in moisture as seen in the water vapour imageries and prevalence of dry weather for 5 days.

v) Withdrawal of SW monsoon is from the southern peninsula and hence from the entire country only after 1st October, when the circulation pattern indicates a change over from the southwesterly wind regime.

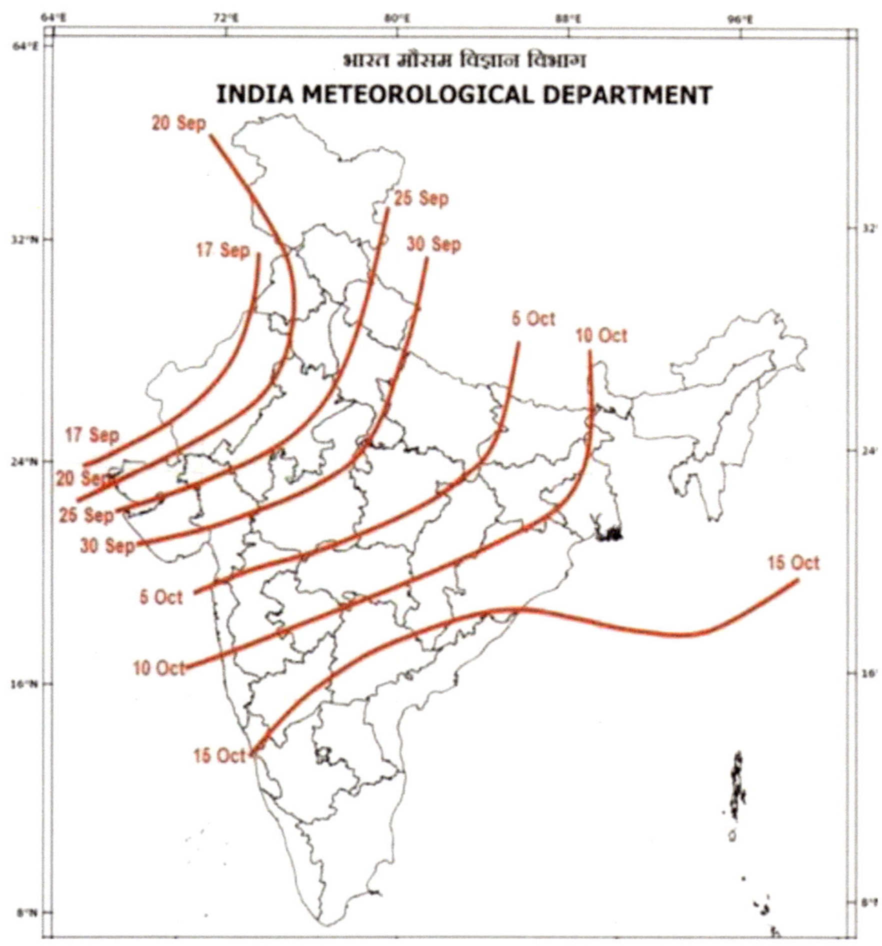

Fig. 13.8: Normal Dates of Withdrawal of SW Monsoon from India

https://internal.imd.gov.in/press_release20200515_pr_804.pdf

The southwest monsoon starts withdrawing from western Rajasthan from September first onwards. When monsoon starts withdrawing, it is called retreating southwest monsoon.

The retreat starts from regions where it reaches last. The retreat of these winds takes place due to weakening of low pressure area in northwestern parts and by 15th October it withdraws in most parts of the Country except from north coastal districts of Andhra Pradesh and Tamilnadu under favorable weather conditions. By end of December it completely withdraws from the Country. The normal dates of withdrawal of southwest monsoon in India can be seen in below Fig. 13.8.

13.8 Breaks in Southwest Monsoon

A break in monsoon period occurs during which the monsoon rains will not occur. The breaks in southwest monsoon may range from few days to few weeks. During the break monsoon period, the monsoon trough shifts more towards the Himalayas resulting in more rainfall near Himalayan region and no rain in the rest of the country.

13.9 Depressions in Southwest Monsoon

The depressions (intense low-pressure system) originating in Arabian sea and Bay of Bengal are responsible for large portion of rainfall during southwest monsoon season. Some depressions may form on land as well. On the average three to four depressions form per month. During the months June and July, depressions in Arabian sea move in northwest or northerly direction producing rainfall in Gujarat and Maharashtra. The absence in depressions lead to failure of rains and leads to droughts particularly in low rainfall regions.

13.10 Significance of Southwest Monsoon Season

i) During the southwest monsoon season, India receives nearly 75% of its annual rainfall.

ii) Southwest monsoon rains contribute of nearly 60% of the water used for irrigation of crops.

iii) A wide range of crops including rice, pearl millet, sorghum, soybean, cotton, groundnut etc. rely on southwest monsoon rainfall.

iv) The ground water recharge also takes place substantially during the southwest monsoon.

13.11 Northeast Monsoon Season

Cold winds blowing from northeastern region of the country during the months of October to December travel over Bay of Bengal, picks up moisture and produces rainfall in Southern most districts of Andhra Pradesh, Tamil Nadu, Kerala Karnataka and Pondicherry. The SWM winds blow from sea to land whereas the northeast monsoon winds blow from land to sea. Tamil Nadu receives nearly 48% of its annual rainfall during northeast monsoon season.

13.12 Cyclones During Northeast Monsoon

i) Tropical cyclones that are most destructive originate in the Indian Ocean particularly in Bay of Bengal during the month of October and November.

ii) The cyclones in the Bay of Bengal form between 8°N to 14°N latitudes. The cyclones move westwards or north-westerly direction to start with and many of them recurve and move in northeast direction.

iii) The coastal regions of Tamil Nadu and Andhra Pradesh are worst hit due to October- November cyclones

13.13 Effect of La Nina on Northeast Monsoon

Conditions associated with La Nino increase rainfall during southwest monsoon but will decrease rainfall during northeast monsoon season.

13.14 Rainfall in North India During Northeast Monsoon Season

During the months of November and December, rainfall occurs in the Gangetic plains and northern states like Delhi, Punjab, and Haryana. But these rains are not due to northeast monsoon. The western disturbances that originate from Iran picks up moisture form Mediterranean Sea produces rainfall in these parts. However the precipitation occurs mostly in the form of snow in Himachal Pradesh, Jammu and Kashmir and Uttarakhand due to movement of western disturbances.

13.15 Rainfall During Summer Season

The rainfall received during the summer season particularly due to thunderstorm activity are pre-monsoon showers. The pre-monsoon thunderstorms occur during the months of April and May particularly in the evenings. In west Bengal and north eastern parts of India, the thunderstorms that produce rainfall are called norwesters. The pre-monsoon thunderstorms received particularly towards the end of May are useful for land preparation to sow crops during Kharif season.

13.16 Madden- Julian Oscillation (MJO)

The Madden-Julian Oscillation is an atmospheric phenomenon which affects the weather pattern across the globe. It causes major fluctuation in tropical weather on weekly to monthly scales. MJO is defined as a disturbance of clouds, wind and pressure moving eastward at a speed of 4-8 meters per second. Sometimes, it takes even 90 days. It is traversing (to move or travel through an area) and is prominent over the Indian and Pacific oceans.

As MJO moves, it splits into two phases called active phase and suppressed phase. The active phase causes increased rainfall due to greater convection in the atmosphere. During the active phase wind converges at the surface level. During suppressed phase, the air descends from the top of the atmosphere downwards and diverges at the surface. As the air descends from higher altitudes, its temperature increases and leads to decrease in rainfall. Whenever the MJO phase is over the Indian ocean during the monsoon season it brings good rainfall in Indian sub-continent. Whenever MJO stays over Pacific Ocean, the monsoon rainfall decreases in India.

14

Climatic Classification of India (Thornthwaite Method)

According to Thornthwaite's method of climatic classification, six major types of climates exist in India (Fig. 14.1) as stated below:

14.1 Perhumid Regions of India (A)

These regions receive rainfall of more than 2500 mm and consist of Konkan region along Western Ghats, Kerala and most parts of northeast India covering Assam, Meghalaya, Arunachal Pradesh and Tripura etc.

14.2 Humid Regions of India (B)

These regions receive annual rainfall ranging from 2000-2500 mm and are found in areas adjoining perhumid climate.

14.3 Moist Sub-Humid Regions (C)

These regions receive rainfall of about 1500 to 2000mm and consist of narrow belt adjoining regions of humid climate of Western Ghats and eastern India consisting of west Bengal and Orissa.

14.4 Dry Sub-Humid Regions (C2)

These regions receive annual rainfall ranging from 800-1500 mm and consist of northern narrow belt of Gangetic plains, eastern parts of Uttar Pradesh, Bihar, Madhya Pradesh, Chhattisgarh, Jharkhand and coastal Andhra Pradesh and parts of Karnataka.

14.5 Semi-Arid Climatic Region (D)

These regions receive annual rainfall ranging from 500- 800 mm and consist of parts of Punjab, Haryana, eastern parts of Rajasthan, Maharashtra, Karnataka and Telangana, western parts of Tamil Nadu and middle Gujarat.

14.6 Arid Climatic Zone (E)

The region receives annual rainfall of less than 500 mm and comprises of western Rajasthan, western Himalayan regions, Kutch region and rain shadow regions of western ghats.

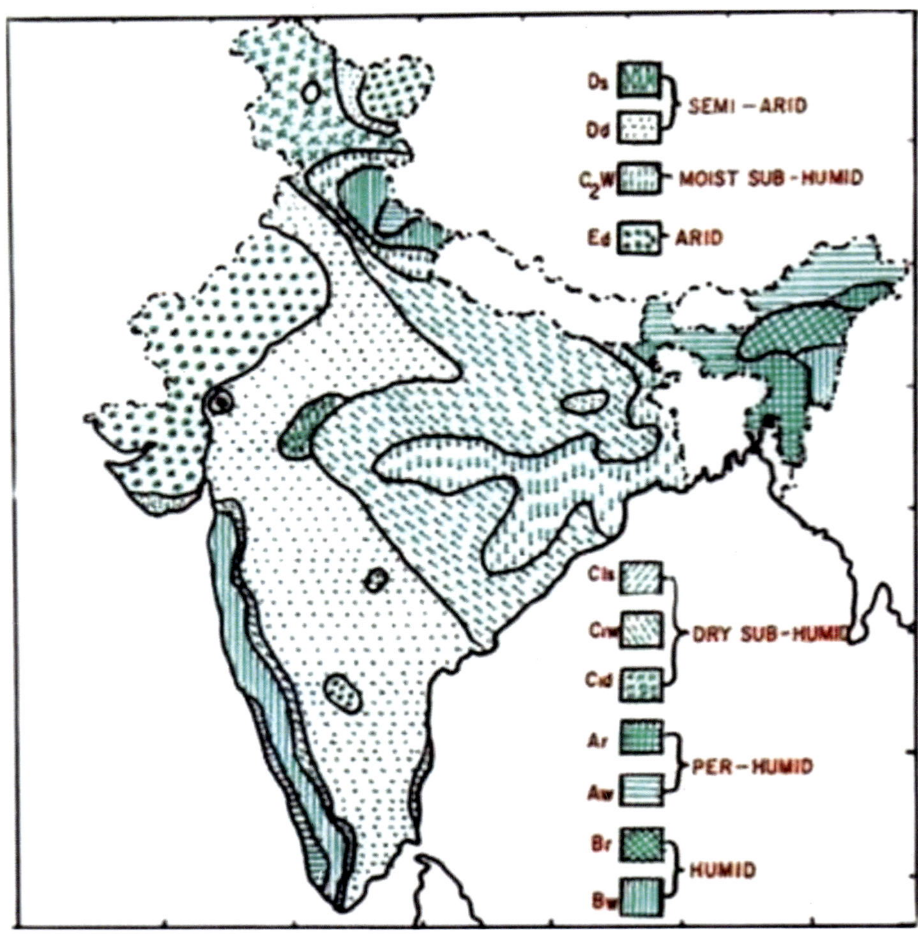

Fig. 14.1: Thornthwaite's Climatic Classification of India
https://targetupsc17.wordpress.com/test-13-ans-6/

15

Climate Change

From the beginning of 18th century, the human activities started increasing gradually. Industrialization, increased use of petroleum products, deforestation, mining and increased use of chemicals have contributed to the pollution of natural resource base. The proportion of gases such as carbon dioxide, nitrous oxide, methane etc started increasing in the atmosphere and more heat is getting trapped in the atmosphere leading to rise in atmospheric temperatures (Fig. 15.1).

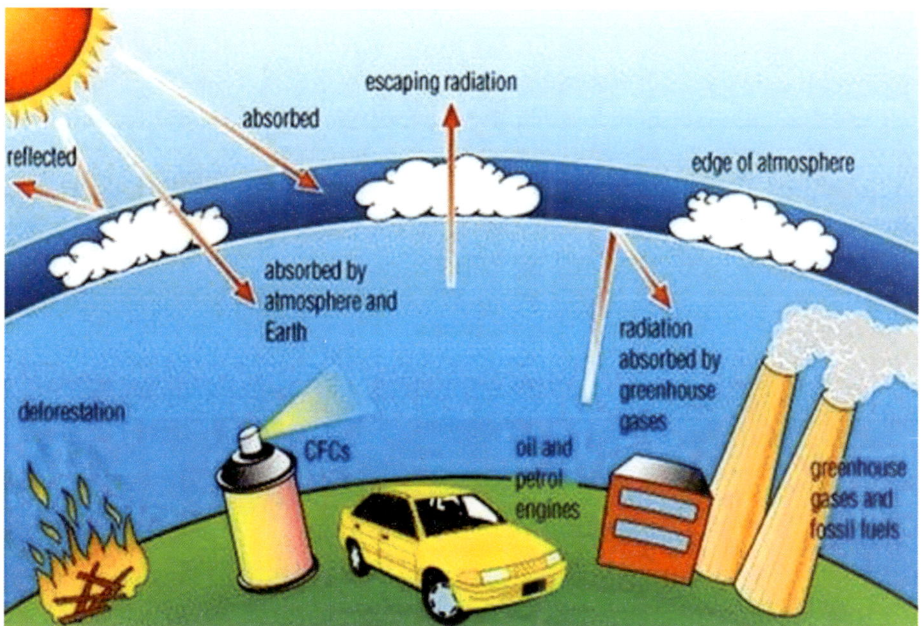

Fig. 15.1: Climate Change Explained
https://www.piercecountywa.gov/7106/Climate-Change-Explained

The changes in atmosphere temperatures have led to changes in weather patterns. This changes in weather pattern triggered extreme situations like heat wave, cold wave, severe cyclonic activity, strong winds and hail storms, floods and droughts of greater intensity creating new records. These changes triggered climatic variability beyond known limits contributing to changing climate.

15.1 Climate Variability: The year-to-year changes in climate within the known limits is called climate variability without any increasing or decreasing trend in the weather parameters over a longer period of time i.e. for several decades.

15.2 Climate Change: The year-to-year changes in climate beyond known limits with either an increasing or decreasing trend is called climate change. The increase in temperatures in the atmosphere will contribute changes in atmospheric pressure, winds, and energy transfer processes in the atmosphere ultimately contributing to changes in weather patterns leading to climate change. The climate change is associated with extreme weather conditions with greater frequency. The climate change is also contributing to greater inter- seasonal as well as intra-seasonal variability of weather.

15.3 Impact of Climate Change on Agriculture in India

The climate change is the major threat for increasing agricultural production to meet the demand for food sufficient for increasing population. The extreme weather events like floods, droughts, heat waves, cold waves, hail storms etc may lead to reduced productivity and even crop failures (Fig. 15.2). In India, rice and wheat alone contribute nearly 240 million tons out of 325 million tons of food grains produced in the country. Both rice and wheat crops were very sensitive to changes in weather. Increasing temperatures during winter season particularly when the wheat crop is in boot formation to milking stage will considerably reduce the yields. Rice crop is known to be very sensitive to maximum temperatures beyond 36°C. The climate change may lead to changes in length of growing season and start of rainy season as a result of changing rainfall patterns. Rise in temperatures will lead to increased demand for water for irrigated crops. More intense rains may lead to increase in runoff and soil erosion. The ground water recharge may be less. Short fall in agricultural production will lead to less supply of agricultural commodities to agro based industries and food processing units. There may be decline in export of agro based products and reduced employment. Further, it may be possible that,

i) The quantity of agricultural produce may deteriorate due to extreme weather conditions.

ii) Frequent hail storms and increased size of hail stones may cause considerable damage to horticultural crops.

iii) Intense storms may contribute to loss of soil fertility.

iv) Increase in temperatures may trigger rapid development of plants, leading to reduced length of crop growth period and hence the crop yields.

v) Under elevated carbon dioxide concentration, crop residues have higher C:N ratio which may reduce decomposition and nutrient supply.

vi) Increased soil temperatures will increase nitrogen mineralization but its availability may decrease due to increased gaseous losses through processes like volatilization and denitrification.

vii) Rise in sea level coupled with increased cyclonic activity may lead to submergence of coastal soils and increase soil salinity.

viii) Increase in temperatures may cause severe heat stress on livestock leading to reduced feed intake, retarded growth and reduced productivity. Acute shortage of fodder and drinking water for cattle cannot be ruled out.

ix) Increasing sea and river water temperatures may affect fish breeding, migration and harvest. Increase in cyclonic activity may affect capture and marketing costs of marine fisheries.

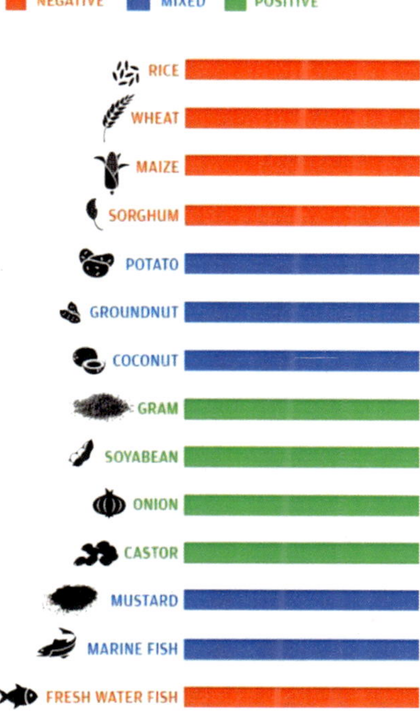

Fig. 15.2: Commodity Wise Impact of Climate Change in India
https://www.downtoearth.org.in/news/agriculture/climate-change-causes-about-1-5-per-cent-loss-in-india-s-gdp-57883

15.4 Role of Human Interventions on Climate Change

The climate change is the result of increased human activities all over the world. The factors that contribute to climate change are

i) Deforestation leading to increased carbon dioxide in the atmosphere.
ii) Increased use of petroleum products leading to increased emission of greenhouse gases.
iii) Over dependence of animal products to meet food requirement and growth of livestock sector.
iv) Urbanisation
v) Increased use of CFC(Chlorofluorocarbon) gases which can deplete ozone in the stratosphere leading to global warming and
vi) Increases in atmospheric methane could be due to greater amount of area devoted to cultivation of paddy and increased number of cattle.

The climate change may not be reversible but can be slowed down through human interventions. The basic principle to combat climate change is through reducing greenhouse gas emissions. It is necessary to

i) Promote afforestation so that the atmospheric carbon dioxide can be trapped by vegetation
ii) Reduced use of fossil fuels
iii) Promote use of renewable sources of energy and reduce thermal power plants.
iv) Reduced use of petroleum products for transportation
v) Minimise consumption of animal products and promote consumption of plant-based products.
vi) Reduced usage of chemical fertilizers.

More intensive research is required to minimize methane emissions particularly in rice cultivation and livestock production. Carbon dioxide is the mostly produced green house gas. Carbon sequestration in the process of capturing and storing atmospheric carbon dioxide. It is one method of reducing Carbon dioxide in the atmosphere with global climate change. Biological carbon sequestration happens when carbon is stored in natural environment. This includes what are known as carbon sinks such as forests, grasslands, soil, oceans and other water bodies. Forests are considered one of the best forms of carbon sequestration. Carbon dioxide is used by plants for photosynthesis and release oxygen to purify the environment. Soil carbon sequestration includes various ways of managing land so that soil absorb and hold more carbon. Increasing soil carbon is accomplished in various ways including.

i) Reducing soil disturbance through minimum tillage.

ii) Changing planting schedules to extend vegetative cover for longer periods.

iii) Controlled grazing by livestock and

iv) Application of compost and recycling plant residues.

15.5 Recent Trends in Climate Change

There are very clear and unambiguous evidences of climate change as witnessed from historical weather data available across different parts of the world.

15.5.1 Global Trends

Scientists from around the world serve as Intergovernmental Panel on Climate Change (IPCC). They found that from the year 1900 to 2020, the world's surface temperature increased by 1.1°C (nearly 2°F) due to burning of fossil fuels that release greenhouse gases like carbon dioxide etc into the atmosphere (Fig. 15.3). This warming is unprecedented in over 2000 years.

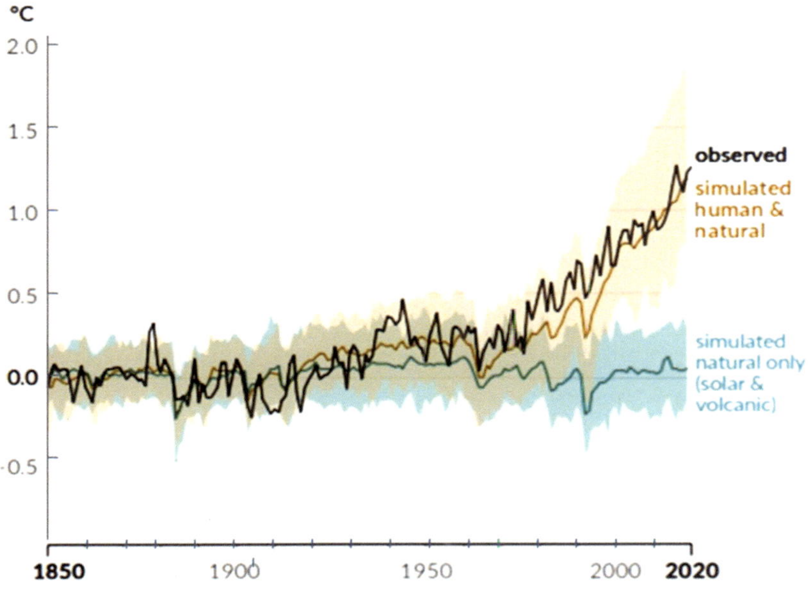

Fig. 15.3: Global Mean Temperature Trends
https://www.ipcc.ch/assessment-report/ar6/

Climate models predict that earth's temperature will rise by 4°C (7.2°F) during 21st century if green house gases continue to rise at present rate without immediate action to reduce greenhouse gas emissions. Models project that it may not be possible to hold global average temperatures to within a change of 1.5°C to 2.0°C. Thus, the amount of climate change by the end of present century depends upon decisions that are made as on now to reduce green house gas emissions. If we do not reduce greenhouse gas emissions, the warming will be about 4.5°C to 5°C.

Further

i) The temperature increases are expected to be greater on land compared to over oceans and

ii) Polar regions and land areas are expected to see large temperature changes.

15.5.2 Climate Change in India

India's average temperature has risen by around 0.7°C during 1901 to 2018. This rise in temperature is largely due to greenhouse gases induced warming. By the end of the 21st century, average temperatures over India are projected to rise by 4.4°C approximately compared to recent past (1976-2005 average) under RCP 8.5 scenario. During the period 1986-2015, temperatures of the warmest day and the coldest night have risen by about 0.63°C and 0.40°C respectively (Fig. 15.4). Further it is projected that

i) The average duration of heat wave events may be doubled and

ii) Combined rise of surface temperature and humidity will amplify heat stress particularly over Indo-Gangetic and Indus River basins.

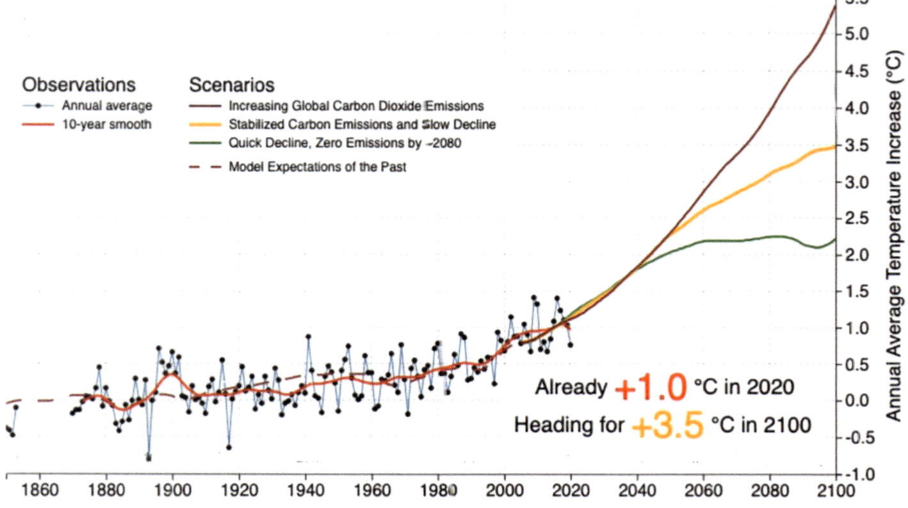

Fig. 15.4: Warming Over India
https://bonpote.com/en/deadly-heatwaves-in-india-and-pakistan-its-only-the-beginning/

15.5.3 Trends that Show Reality of Climate Change

Scientists say earth has experienced cycles of warming and cooling over the ages. But the significant warming trend we are experiencing now is different from the past. The difference is mainly due to increased human activity. The scientists have identified nine significant trends to show that climate is changing during recent years.

i) **Rising sea levels:** The global sea levels rose by about 8 inches over the last 100 years according to NASA (National Aeronautics and Space Administration). It is estimated that the rise in sea level may be at least 12 inches above 2000 levels by the year 2100. The rise is attributed to melting of glaciers, ice sheets and expansion of water due to rising temperature.

ii) **Rising temperatures:** The average temperature of the earth's surface has increased by 2°F since the late 19th century according to NOAA (National Oceanic and Atmospheric Administration). The IPCC which includes more than 1300 scientists from all over the world predict a temperature rise of 2.5 to 10°F over next 100 years (Fig. 15.5).

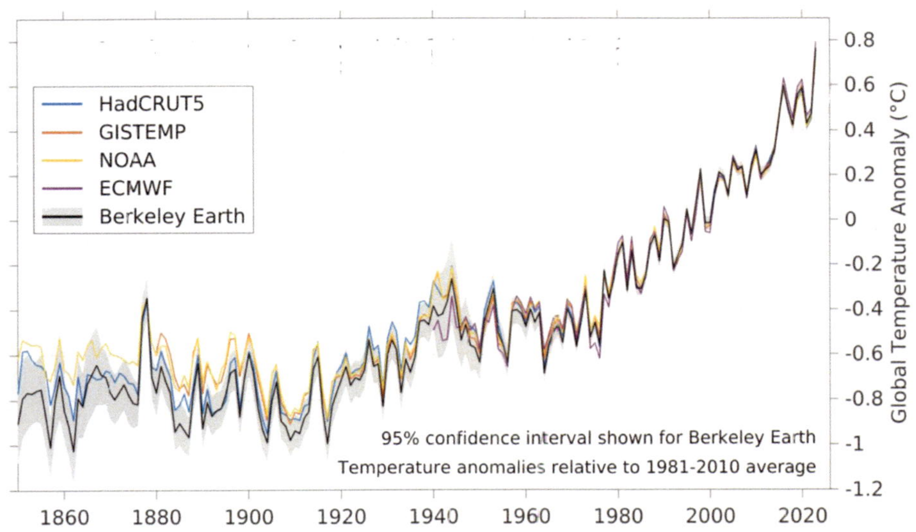

Fig. 15.5: Global Warming Trends
https://berkeleyearth.org/global-temperature-report-for-2023/

iii) **Oceans are warming:** Oceans absorb more than 90 per cent of the increased atmospheric heat associated with greenhouse gas emissions caused by human activity. According to NASA, the top 328 ft of the ocean's surface has warmed 0.6°F since 1969 (Fig. 15.6).

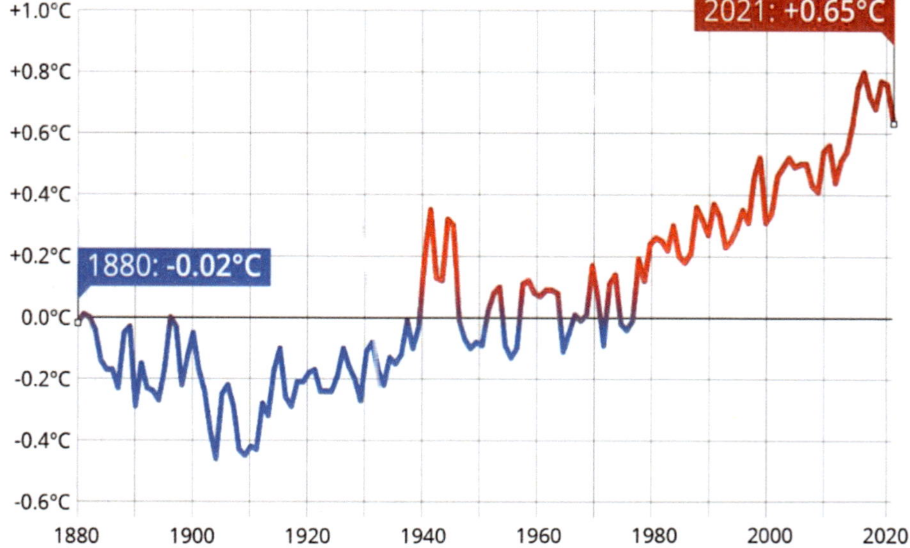

Fig. 15.6: Ocean Warming
https://www.statista.com/chart/19418/divergence-of-ocean-temperatures-from-20th-century-average/

iv) **Shrinking of Ice sheets:** Data from NASA's gravity recovery and climate experiment shows Greenland lost an average of about 280 billion tons of ice per year between 1993 and 2019 while Antarctica lost 150 billion tons per year.

v) **Increase in frequency of extreme weather events:** There is considerable increase in extreme weather events like heat waves, cold waves, droughts, hail storm and severe cyclones in the recent past.

vi) **Increase in relative humidity:** According to NOAA, measurements taken from weather stations are showing increased humidity due to the amount of water vapour in the air (Fig. 15.7).

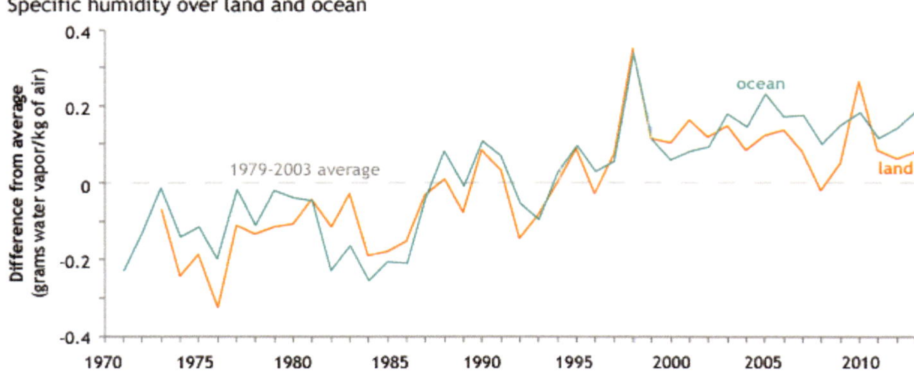

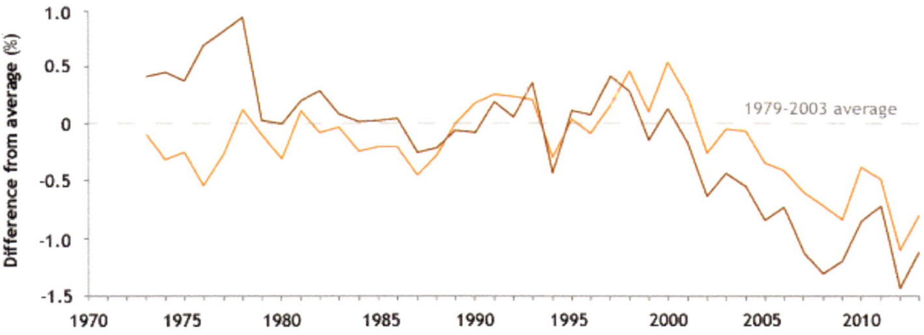

Fig. 15.7: Global Change in Humidity
https://www.climate.gov/news-features/understanding-climate/2013-state-climate-humidity

vii) **Reduction in snow cover:** Satellite data shows that the amount of spring snow cover in the Northern hemisphere has decreased (Fig. 15.8) over the past five decades and the snow melting earlier.

Fig. 15.8: Reduction in Snow Cover
https://ugc.berkeley.edu/background-content/snow-ice-cover/

viii) **Retreating glaciers:** Glaciers in the US and the around the world has generally shrunk since 1960's. The rate of which glaciers are melting has accelerated over the last decade (Fig. 15.9).

Fig. 15.9: Retreating Glaciers
https://eos.org/articles/a-future-of-retreating-glaciers-in-the-himalayas

ix) **Oceans are becoming acidic:** According to NOAA, the acidity of the surface ocean waters has increased by 30 percent since beginning of the industrial revolution.

All the above evidences prove beyond doubt that climate change is a reality and there is urgent need to take adequate measures to prevent global climate change by adopting technologies that can minimize greenhouse gas emissions without any further loss of time.

15.5.4 Depletion of Ozone and Ozone Hole

The ozone present in the stratosphere absorbs harmful ultra-violet radiation coming from the sun and is protecting life on earth. Ozone layer depletion is gradual thinning of ozone layer in the stratosphere (Fig. 15.10). It is caused due to release of chemical compounds like gaseous bromine and chlorine into the atmosphere from industries and other human activities.

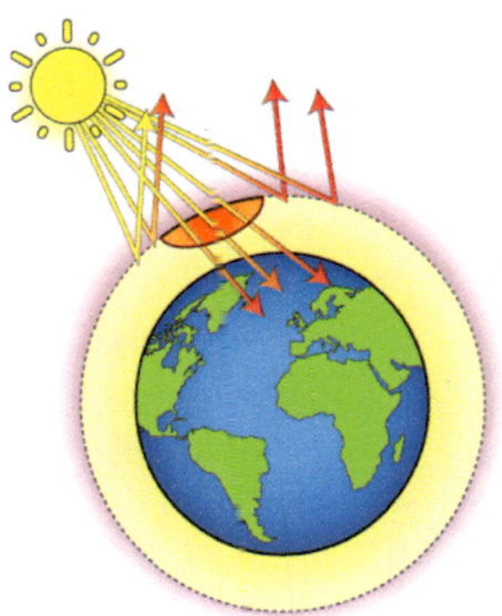

Fig. 15.10: Depletion of Ozone
https://byjus.com/biology/ozone-layer-depletion/

When chlorine and bromine atoms come in contact with ozone, ozone molecules get destroyed. The ozone depleting substances that contain chlorine include chlorofluorocarbon, carbon tetrachloride, hydrofluoro carbon and methyl chloroform. The ozone depleting substances that contain bromine are halons, methyl bromide, hydrofluorobromo carbons. The nitrogenous compounds such as nitrous oxide, nitrogen dioxide also contribute to depletion of ozone.

15.5.5 Ozone hole

The thinning of ozone layer over Antarctica came to be known as Ozone hole. In the southern hemisphere, the South Pole is part of very large land mass (Antarctica) that is completely surrounded by ocean. The conditions allow formation of a very cold region in the stratosphere during winter over Antarctica continent. Further strong winds will be circulating around the edge of the Antarctica region (Fig. 15.11). The very low stratospheric temperatures lead to formation of clouds that are responsible for chemical changes leading to production of chemically active chlorine and bromine results in rapid ozone depletion when sunlight returns to Antarctica during the months of September and October each year. This results in formation of ozone hole. Similarly conditions do not exist in Arctic region of Northern hemisphere as the winter time temperatures in the Arctic stratosphere are not persistently low as observed in Antarctica region.

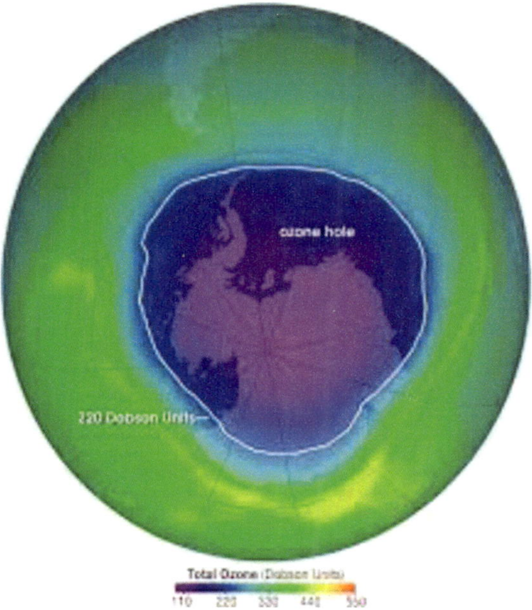

Fig. 15.11: Ozone Hole
https://ozonewatch.gsfc.nasa.gov/facts/hole_SH.html

16

Some Physical Parameters and Their Units

In meteorology, we often use some of the physical parameters to explain the physical processes taking place in the atmosphere. These parameters must have already appeared in the earlier chapters. However, these are given in this chapter to enable the readers to get acquainted once again.

16.1 Momentum

Momentum is defined as the product of mass and its velocity. In S.I. system units mass is expressed in kilograms and velocity in meters per second.

Momentum = Mass x Velocity

Units for momentum (S.I. system) = kg. m/sec

In C.G.S. System, mass is expressed in grams and velocity in cm/sec.

Units for momentum (C.G.S System) = gm.cm/sec

16.2 Force

The force is defined as the product of mass and its acceleration.

F= ma

Where F = Force

m= mass of the body

a= acceleration

The S.I. unit of force F= (Kg) (m/sec^2)

= Kg.m/sec^2

The S.I. unit for force is also called Newton, if a body of mass 1 Kg is moving with an acceleration of 1 m/sec^2, the force is equal to 1 Newton.

In C.G.S. System, the unit of force is Dyne. If a body of mass 1gm is moving with an acceleration of 1cm/sec^2, then the force is equal to one dyne.

One Newton = 10^5 dynes.

16.3 Work

Work is said to be done when a force moves a body in the direction of the force

Work= force x distance through which the body moves in the direction of the force.

W = F x s

Where W = Work done

F = Force

s = distance through which the body moves

The S.I. unit of work is Joule (J).

Joule is defined as work done by a force of one Newton causing a displacement of the body by one meter.

The C.G.S. unit of work is one erg.

The work done is one erg of a force of one dyne displaces a body by 1cm.

One Joule = 10^7 ergs.

16.4 Power

The power is defined as the rate at which work is done or works done for one second time.

The S.I. unit of Power = Joule/sec = Watt

The C.G.S. unit of Power = Ergs/sec

The S.I. unit of Power is Watt and one Watt is equal to one Joule of work done in one second.

The Power is expressed as horse power (HP) in F.P.S. units.

1H.P. = 745.7 Watts

16.5 Surface Tension

The surface tension of a liquid is mainly a force that acts to reduce the surface area of the liquid. The directed contracting force which attracts the molecules at the surface of a liquid towards the interior of the liquid is called surface tension.

The S.I. unit for Surface tension = Newton/meter

The C.G.S. unit for Surface tension = Dyne/cm

16.6 Pressure

The pressure is defined as the force per unit area.

The S.I. unit of Power is Newton per square meter (Newton/m^2).

The C.G.S. unit of Power is dyne per square centimeter (dyne/cm^2).

The units used to express atmospheric pressure were given earlier.

16.7 Light Intensity

The fundamental unit of light is the Candela. It indicates the light given by one candle. To be very precise, it is equivalent to monochromatic radiation of frequency 540 x 1012 hertz and that has a radiant intensity in that direction of 1/683 Watt per steradian.

Monochromatic light is a single wavelength light.

Steradian is the SI unit of solid angle. It is equal to the angle at the center of a sphere Subtended by a part of the surface of the sphere equal to the square of the radius.

One Candela = 12.57 Lumens

One lumen is approximately the light of one candle shining on 1sq. feet area at a distance of 1ft away from the source.

To measure light intensity received at a surface, the most commonly used unit is Lux.

One lux = 1 lumen/sq. meter

One phot = 10,000 Lux.

The Phot is a photometric unit of Illuminance.

Bibliography

Balasubramanian, T.N., Jagannathan, R. and Geethalaxmi, V. 2021. Agro-Climatology Advances and Challenges. New India Publishing Agency (NIPA). ISBN: 9788194766896. Pp 332.

Bishnoi, O.P. 2007. Principles of Agricultural Meteorology. Oxford Book Company. ISBN -13: 9788189473013. Pp 420.

Buchan, A. 2023. Introductory Text-Book of Meteorology. Legare Street Press.ISBN-13: 978-1020278716. Pp 256.

Byers, H.R. 1974. General Meteorology. New York (McGraw-Hill), 4th Edition. Pp. 461.

Chandrasekhar, A. 2010. Basics of Atmospheric Science. PHI Learning Private Limited, New Delhi. ISBN: 13-9789391818241. Pp.464.

Chang, J.H. 1968. Climate and Agriculture–An Ecological Survey. Aldine Publishing Company. ISBN: 9781351527972, 13-51527975. Pp. 320.

Critchfield, H.J. 2008. General Climatology. Pearson Education India. ISBN: 13-9332555242-978. Pp 385.

Das, P.K. 2018. The Monsoons. National Book Trust. ISBN-10: 81237--11239. Pp 254.

Donald, C. A. and Robert H.. 2021. Meteorology: An Introduction to Weather, Climate, and the Environment (13th Edition). Brooks/Cole Publisher. ISBN-13: 9780357452073. Pp 736.

Donald, C. A. and Robert, H. 2017. Essentials of Meteorology. Cengage Learning Publisher. ISBN-13: 9781305628458. Pp 508.

Franklyn, W. C. 1980. Introduction to Meteorology (3rd Edition). John Wiley and Sons. ISBN-13: 978-0471164814. Pp 505.

Frederick. J.E. 2007. Principles of Atmospheric Science. Jones and Bartlett Publishers, Inc. ISBN-13: 9780763740894. Pp 211.

Galvin, J. F. P. 2016. An Introduction to the Meteorology and Climate of the Tropics. Wiley-Blackwell eBook. ISBN: 13-9781119086239. Pp 465.

Ghadekar S.R. 2002. Practical Meteorology: Data Acquisition Techniques and Instruments, Agromet Publishers Nagpur. Pp 68.

Ghadekar S.R. 2016. Meteorology (14th Edition),Agromet Publication, Nagpur.

Gill, A. 1982. Atmosphere-Ocean- Dynamic Meteorology. Academic Press; 1st Edition (13 December 1982). ISBN-13: 9780122835223. Pp 680.

Glickman, T.G. 2000. Glossary of Meteorology (2nd Edition).University of Chicago Press. ISBN-13: 9781878220349. Pp 850.

Houghton, H. G. 1985. Physical Meteorology. Cambridge, Mass. MIT Press. ISBN: 0262081466, 9780262081467. Pp. 430.

IPCC, 2014: Climate Change 2014: Synthesis Report. Contribution ofWorking Groups I, II & III to the Fifth Assessment Report of the Intergovernmental Panel on Climate Change [Core Writing Team, R.K. Pachauri and L.A. Meyer (eds.)]. IPCC, Geneva, Switzerland, 151 pp.

James, R. H. and Gregory J. H. 2013. An Introduction to Dynamic Meteorology. Academic Press Elsevier. ISBN: 13-9780123848666. Pp 519

Kakde, J.R. 1985. Agricultural Climatology. Metropolitan Book. Co. ISBN-13: 9788120000056. Pp 387.

Kelkar, R.R. 2010. Climate Change-A holistic view. BS Publications, Hyderabad. ISBN-13: 9788178002491. Pp 208.

Khan, N. and Dubey, D.V. 2021. Climate Science. Bhavya Books (BET) Tm New Delhi. ISBN-13: 978-9383992508. Pp 276.

King, F.H. 2018. The Text Book of Physics of Agriculture. Forgotten Books. ISBN: 13-9781033217290. Pp 604.

Lal, D.S. 2010. Climatology. Sharda Pustak Bhawan. ISBN: 10-8186204121. Pp 231.

Lenka, D. 1998. Climate, Weather and Crops in India. Kalyani Publishers. ISBN: 9788170969846. Pp. 490.

Lilly, D.K. and Chen, T.G. 2010. Introduction to Mesoscale Meteorology- Theories, Observation and Models. Springer; SBIN: 13-978-9048183906. Pp.781.

Mahi, G.S. and Kingra, P.K. 2014. Fundamentals of Agrometeorology. Kalyani Publishers. ISBN-13: 9789327235487. Pp 392.

Mavi, H.S. and Tupper J.C. 2004. Principles and Applications of Climate Studies in Agriculture. CRC Press. ISBN-10: 1560229721. Pp 351.

Mavi, H.S. 1994. Introduction to Agrometeorology. Oxford & IBH Publishing Co. Pvt Ltd. New Delhi. ISBN-10: 8120409 08. Pp 281.

Murthy, B.P. 2004. Environmental Meteorology. I.K.International (Pvt.) Ltd. ISBN-10: 8188237108. ISBN-13: 9788188237104. Pp 278.

Rakhecha, P.R. and Singh, V.P. 2009. Applied Hydrometeorology. Capital Publishing Company, New Delhi, Kolkata. ISBN-10: 8 85589631. Pp 384.

Rao, G.S.L.H.V. P. 2008. Agricultural Meteorology. Prentic-Hall of India (P) Ltd., ISBN-13: 978812033338. Pp 285.

Rao, G.S.L.H.V.P., Rao V.U.M. and Rao, D.V.S. 2019. Climate Change and Agriculture. New India Publishing Agency. ISBN-13: 978-9387973626. Pp 555.

Rose, C.W. 2013. Agricultural Physics. Pergamon Press Ltd. Oxford. ISBN- 13: 9781483139258, 10: 1483139255. Pp.248.

Rosenberg, N.J. 1974. Microclimate: The Biological Environment. Wiley. ISBN-10: 0471736155. Pp 315.

Rosenberg, N.J.; Blad, B.L. and Verma, S.B.1983. Microclimate - The Biological Environment. Wiley. ISBN-13: 978-0471060666.Pp 495.

Shonk, J. 2013. Introducing Meteorology: A Guide to the Weather (Introducing Earth and Environmental Sciences). Dunedin Academic Press. ISBN-13: 9781780460024. Pp. 160.

Spiridonov, V. and Ćurić, M. 2020. Fundamentals of Meteorology. Springer International Publishing. ISBN-13: 9783030526542. Pp 437+XXV.

Srivastava, G.P. 2008. Surface Meteorological Instruments and Measurement Practices. Atlantic Publishers New Delhi. ISBN-13: 97881269-09681. Pp 464.

Steven, A. A. and Knox. J.A. 2011. Meteorology (3rd Edition).Jones & Bartlett Publishing. ISBN-13: 9781449631758. Pp 580.

Stevermer, A. J. 2002. Recent Advances in and issues in Meteorology. ISBN: 9781573563017. Pp 278.

Trewartha, G.T. and Lyle H. H. 1981. An Introduction to Climate, McGraw Hill International Book Company. ISBN-10: 0070651523, ISBN-13: 978-0070651524 Pp. 407.

Varshneya, M.C. and Pillai, P.B. 2003. Text Book of Agricultural Meteorology. ICAR.ISBN-13 : 9788171640195. Pp 221.

Wallace, J.M. and Hobbs. P.V. 2006. Atmospheric Science: An Introductory Survey. Academic Press. ISBN-13: 9780127329512. Pp 483.

WMO. 2013. Guidelines for Trainers in Meteorological, Hydrological and Climate Services. ISBN-13: 9789263111142. Pp 84.

Wyngaard, J.C. 2010. Turbulence in the Atmosphere. Cambridge University Press. ISBN: 13-9780521887694. Pp 357.

Web Resources

http://apes2k15-earthsystemsandresources.weebly.com/composition-and-structure.html
https://berkeleyearth.org/global-temperature-report-for-2023/
https://bonpote.com/en/deadly-heatwaves-in-india-and-pakistan-its-only-the-beginning/
https://byjus.com/biology/ozone-layer-depletion/
https://commons.wikimedia.org/wiki/File:Cyclone-anti-cyclone-wind-direction-air-pressure.jpg
https://commons.wikimedia.org/wiki/File:Hurricanes,_cyclones,_and_typhoons.jpg
https://depositphotos.com/photos/dew-on-leaves.html
https://dictionary.cambridge.org/dictionary/english/hail
https://digitallylearn.com/thunderstorms-and-tornadoes-upsc-ias/
https://dokumen.tips/documents/india-wind-zone.html
https://drrajkumars.com/atmospheric-pressure/
https://en.wikipedia.org/wiki/Cumulus_cloud
https://en.wikipedia.org/wiki/Mist
https://eos.org/articles/a-future-of-retreating-glaciers-in-the-himalayas
https://eschooltoday.com/learn/the-atmosphere/
https://future-seafarer.com/col/
https://geography4u.com/climate-zones-in-india/
https://internal.imd.gov.in/press_release20200515_pr_804.pdf
https://link.springer.com/chapter/10.1007/978-3-030-22403-5_1
https://mausam.imd.gov.in/
https://ozonewatch.gsfc.nasa.gov/facts/hole_SH.html
https://prepp.in/news/e-492-seasons-in-india-geography-notes
https://quizlet.com/788076711/weather-and-climate-exam-2-flash-cards/
https://recursos.edu.xunta.gal/sites/default/files/recurso/1605879478/importance_of_the_atmosphere.html
https://salepeaket.live/product_details/8906348.html
https://scied.ucar.edu/learning-zone/how-weather-works/weather-fronts
https://targetupsc17.wordpress.com/test-13-ans-6/
https://testbook.com/ias-preparation/pressure-belts-of-the-earth
https://ugc.berkeley.edu/background-content/snow-ice-cover/
https://www.britannica.com/biography/Aristotle
https://www.britannica.com/science/cyclostrophic-wind
https://www.britannica.com/science/phase-state-of-matter
https://www.buddinggeographers.com/global-climate-vulnerability-and-resilience/
https://www.chem.fsu.edu/chemlab/chm1045/estructure.html
https://www.civilsdaily.com/the-winter-season-january-february/
https://www.climate.gov/news-features/understanding-climate/2013-state-climate-humidity
https://www.cropsmart.com.au/should-i-spray-after-a-frost/
https://www.downtoearth.org.in/news/agriculture/climate-change-causes-about-1-5-per-cent-loss-in-india-s-gdp-57883

https://www.geeksforgeeks.org/rainfall-in-india/
https://www.geeksforgeeks.org/western-disturbances-in-india/
https://www.hindustantimes.com/india-news/fires-dust-weather-plunge-capital-into-air-emergency-101667323926625.html
https://www.insightsonindia.com/world-geography/physical-geography-of-the-world/climatology/world-climatic-regions/classification-of-climate-of-the-world/
https://www.ipcc.ch/assessment-report/ar6/
https://www.ksla.com/2018/09/10/weather-or-not-how-does-fog-form/
https://www.mapsofindia.com/maps/india/annualrainfall.htm
https://www.metlink.org/resource/in-depth-the-global-atmospheric-circulation/
https://www.metlink.org/resource/in-depth-the-global-atmospheric-circulation/
https://www.moldpres.md/en/news/2020/03/04/20001914
https://www.ncmrwf.gov.in/
https://www.noaa.gov/jetstream/atmosphere/transfer-of-heat-energy
https://www.piercecountywa.gov/7106/Climate-Change-Explained
https://www.quora.com/What-is-the-relation-between-agriculture-and-weather
https://www.researchgate.net/publication/225458262_Meteorological_features_at_6523_m_of_Mt_Qomolangma_Everest_between_1_May_and_22_July_2005
https://www.researchgate.net/publication/226241280_Sufferer_and_cause_Indian_livestock_and_climate_change/figures?lo=1
https://www.researchgate.net/publication/253640760_Alexander_von_Humboldt%27s_charts_of_the_Earth%27s_magnetic_field_an_assessment_based_on_modern_models/figures?lo=1
https://www.researchgate.net/publication/262932181_Enhancing_crop_resilience_to_combined_abiotic_and_biotic_stress_through_the_dissection_of_physiological_and_molecular_crosstalk/figures?lo=1
https://www.researchgate.net/publication/285020645_Nationwide_classification_of_forest_types_of_India_using_remote_sensing_and_GIS/figures?lo=1
https://www.researchgate.net/publication/287807359_Infrared_and_Skin_Friend_or_Foe
https://www.researchgate.net/publication/321532699_A_Text_Book_on_Agricultural_Meteorology
https://www.researchgate.net/publication/329138708_Understanding_seismic_body_waves_retrieved_from_noise_correlations_Toward_a_passive_deep_Earth_imaging/figures?lo=1
https://www.sciencefacts.net/plancks-law.html
https://www.sciencelearn.org.nz/images/240-vertical-structure-of-the-atmosphere
https://www.science-sparks.com/what-is-a-sea-breeze/
https://www.slideshare.net/HarshitaPant6/indian-ocean-dipole-iod
https://www.solpass.org/science6-8-new/standards/standard_6.3.html
https://www.statista.com/chart/19418/divergence-of-ocean-temperatures-from-20th-century-average/
https://www.techtarget.com/whatis/definition/heat
https://www.tropmet.res.in/
https://www.zigya.com/question/UjBWRlRqRXhNREEzT1RjMA==

Index